Philosophy
DeMYSTiFieD®

DeMYSTiFieD® Series

Accounting Demystified
Advanced Calculus Demystified
Advanced Physics Demystified
Advanced Statistics Demystified
Algebra Demystified
Alternative Energy Demystified
Anatomy Demystified
asp.net 2.0 Demystified
Astronomy Demystified
Audio Demystified
Biology Demystified
Biotechnology Demystified
Business Calculus Demystified
Business Math Demystified
Business Statistics Demystified
C++ Demystified
Calculus Demystified
Chemistry Demystified
Circuit Analysis Demystified
College Algebra Demystified
Corporate Finance Demystified
Databases Demystified
Data Structures Demystified
Differential Equations Demystified
Digital Electronics Demystified
Earth Science Demystified
Electricity Demystified
Electronics Demystified
Engineering Statistics Demystified
Environmental Science Demystified
Everyday Math Demystified
Fertility Demystified
Financial Planning Demystified
Forensics Demystified
French Demystified
Genetics Demystified
Geometry Demystified
German Demystified
Home Networking Demystified
Investing Demystified
Italian Demystified
Java Demystified
JavaScript Demystified
Lean Six Sigma Demystified
Linear Algebra Demystified

Logic Demystified
Macroeconomics Demystified
Management Accounting Demystified
Math Proofs Demystified
Math Word Problems Demystified
MATLAB® Demystified
Medical Billing and Coding Demystified
Medical Terminology Demystified
Meteorology Demystified
Microbiology Demystified
Microeconomics Demystified
Nanotechnology Demystified
Nurse Management Demystified
OOP Demystified
Options Demystified
Organic Chemistry Demystified
Personal Computing Demystified
Philosophy Demsytified
Pharmacology Demystified
Physics Demystified
Physiology Demystified
Pre-Algebra Demystified
Precalculus Demystified
Probability Demystified
Project Management Demystified
Psychology Demystified
Quality Management Demystified
Quantum Mechanics Demystified
Real Estate Math Demystified
Relativity Demystified
Robotics Demystified
Sales Management Demystified
Signals and Systems Demystified
Six Sigma Demystified
Spanish Demystified
sql Demystified
Statics and Dynamics Demystified
Statistics Demystified
Technical Analysis Demystified
Technical Math Demystified
Trigonometry Demystified
uml Demystified
Visual Basic 2005 Demystified
Visual C# 2005 Demystified
xml Demystified

Philosophy
DeMYSTiFieD®

Robert Arp

Jamie Carlin Watson

New York   Chicago   San Francisco   Lisbon   London   Madrid   Mexico City
Milan   New Delhi   San Juan   Seoul   Singapore   Sydney   Toronto

The McGraw·Hill Companies

Cataloging-in-Publication Data is on file with the Library of Congress

McGraw-Hill books are available at special quantity discounts to use as premiums and sales promotions, or for use in corporate training programs. To contact a representative please e-mail us at bulksales@mcgraw-hill.com.

Philosophy DeMYSTiFieD®

1 2 3 4 5 6 7 8 9 0   DOC/DOC   1 9 8 7 6 5 4 3 2 1

ISBN 978-0-07-171766-3
MHID 0-07-171766-8

Sponsoring Editor
Judy Bass

Acquisitions Coordinator
Michael Mulcahy

Editing Supervisor
David E. Fogarty

Project Manager
Ranjit Kaur,
Glyph International

Copy Editor
Bev Weiler

Proofreader
Eina Malik,
Glyph International

Production Supervisor
Richard Ruzycka

Composition
Glyph International

Art Director, Cover
Jeff Weeks

Cover Illustration
Lance Lekander

About the Authors

Robert Arp, Ph.D., is a philosopher and semantic information architect at OntoReason, LLC, a software and informatics company.

Jamie Carlin Watson, Ph.D., is Assistant Professor of Philosophy and Chair of the Department of Religion and Philosophy at Young Harris College.

Contents

	Introduction		*xi*
Part I	**Philosophy and Correct Reasoning**		**1**
	CHAPTER 1	**An Overview of Western Philosophy**	**3**
		What Is Philosophy?	4
		Domains of Philosophy	8
		The Beginnings of Philosophy	11
		Philosophical Eras	16
		The Value of Philosophy	17
		Quiz	20
	CHAPTER 2	**Thinking Critically about Reality**	**23**
		Claims	24
		Arguments and Evidence	25
		Argument Strategies	32
		Fallacies	41
		Quiz	49
	CHAPTER 3	**Knowledge and the Problem of Skepticism**	**51**
		What Does It Mean to "Know"?	52
		An Obstacle to Knowing: Skepticism	54
		Four Arguments for Skepticism	58
		Quiz	67
	CHAPTER 4	**Responses to Skepticism**	**69**
		René Descartes to the Rescue	70
		The "*Cogito*" Argument	72
		Descartes Gets His Senses Back	75

John Locke Breaks the Rules   79
Empiricism Becomes a Plausible Philosophical View   80
Quiz   86

**CHAPTER 5   The Problem of Induction and the Development
of Externalism   89**
David Hume Revives Skepticism about Science   90
Attempts to Solve the Problem of Induction   93
Hume's Aftermath: From Kant to the Twentieth Century   99
Immanuel Kant: Awakening a Giant   100
Logical Positivism: Back to the Senses   102
Edmund Gettier: A Change in the Curriculum   103
The Move to Externalism   105
Quiz   106

CHAPTER 6   The Mind-Body Problem   109
What Is a "Mind"?   110
Qualities and Types of Mind   110
Ancient Western Mythological Conceptions of Mind   114
Ancient Greek Conceptions of Mind   115
Descartes's Argument for Substance Dualism   116
A Contemporary Argument for Property Dualism   118
The Mind-Body Problem (or Problems)   119
The Staying Power of Consciousness   130
Quiz   131

CHAPTER 7   Personhood and Personal Identity over Time   133
Are You a Person?   134
Criteria for Personhood   134
Criterion 1: The Capacity for Reason   139
Criteria 2 and 3: Mental States and Language   141
Are You the Same "You" through Time?   145
Is Your Mind "You"?   147
Conscious Minds in Bodies   147
"Memories... Like the Corners of My Mind"   149
Are Bodies or Minds Necessary or Sufficient for
Personal Identity?   151
Quiz   153

CHAPTER 8   Freedom and Determinism   155
Different Types of Freedom   156
Alternate Possibilities Freedom (APF) and
Moral Responsibility   159
A Deterministic Universe   161

Why Do I Act Like I Do? Nature, Nurture, and Motivation   164
Hard Determinism   165
Hard Indeterminism   166
Compatibilism   168
Are You Even Free to Read On?   171
Quiz   172

CHAPTER 9   The Question of God's Existence   175
The Concept of "God"   176
Philosophical Approaches to God's Existence   177
Arguments for God's Existence   181
An Argument against God's Existence:
   The Problem of Evil   198
Quiz   203

CHAPTER 10   Moral Philosophy   205
What Is "Moral Philosophy"?   207
How Is a Moral Decision Different from
   Any Other Decision?   208
Normativity   209
Moral Theories   211
Quiz   222

CHAPTER 11   Moral Decision-Making   225
How to Evaluate Moral Claims   226
Two Problematic Moral Theories   235
Quiz   240

CHAPTER 12   Justice, Rights, and Government   243
What Are Groups of Humans Like?   244
Paternal Governments   246
Liberal Governments   247
Freedom and Ownership   249
Law and Morality   252
Individual Rights vs. the Common Good   254
Justice   258
Can't We Just All Get Along?   261
Quiz   262

Final Exam   265
Answers to Quizzes and Final Exam   285
Suggested Reading   287
Index   293

Introduction

Philosophy is everywhere. Philosophical questions and reasoning strategies underwrite every field of inquiry, from chemistry to politics. Yet, philosophy is difficult, so we have written this book to help you ease into the complex and abstract discussions among philosophers. After spending time with this book, you should have a general sense of what philosophy is all about, and you may be able to be read a few of the primary sources of philosophy; that is, you should be able to reap the rewards of reading Plato, Thomas Aquinas, and Immanuel Kant.

How to Use This Book

This book is divided into five sections and twelve chapters. At the end of each chapter is a quiz. The quizzes are "open-book," and you should refer to the chapter texts when taking them. Stick with a chapter until you get most of the answers right. Also, at the end of the book there is a final exam. The final exam contains questions drawn from the chapters. Take this exam when you have finished all the chapter quizzes. A satisfactory score is at least 75 percent correct answers. The answers to the chapter quizzes and the final exam are listed in the back of the book.

We recommend that you complete one chapter a week. An hour or two daily ought to be enough time for this. Don't rush yourself; give your mind time to absorb the material. But don't go too slowly either. Take it at a steady pace and keep it up. That way, you'll complete the course in a few months. (As much as we all wish otherwise, there is no substitute for "good study habits.") When you're done with the course, you can use this book, with its comprehensive index, as a permanent reference. All the best to you!

Robert Arp and Jamie Watson

Philosophy
DeMYSTiFieD®

Part I

Philosophy and Correct Reasoning

chapter 1

An Overview of Western Philosophy

What does it mean to "study philosophy"? Where did the concept of "philosophy" come from? What motivated it? Understanding the goals of philosophers and what motivated these goals is the key to understanding the sorts of questions they ask and their proposed methods of answering them. In this chapter, we will offer a brief sketch of the motives of the earliest philosophers, the nature of their philosophical projects, and the Western tradition they inaugurated.

CHAPTER OBJECTIVES

In this chapter, you'll learn about...

- The nature of Western philosophy
- The branches of Western philosophy
- The beginnings of Western philosophy
- Philosophical eras

What Is Philosophy?

Philosophy is the systematic study of reality using good reasoning in order to clarify difficult questions, solve significant problems, and enrich human lives. This is a standard conception of the tradition of Western philosophy that began with Greeks such as Thales, Anaximander, and Anaximenes, all of the city of Miletus on the coast of what is now Turkey. Eastern philosophy, which has an equally ancient and rich tradition, would also accept much of this conception. Eastern philosophers are also interested in the study of objective reality, though not so much for its own sake.

A primary difference between Eastern and Western approaches to philosophy is that Eastern philosophers do not as firmly distinguish their broadly religious perspectives from their philosophical investigations. Western philosophers are known for drawing a sharp distinction between their study of reality and their religious convictions. Eastern philosophers are, traditionally, more comfortable allowing their religious beliefs to motivate and inform their investigations toward a broadly religious end, e.g., nirvana. Though Eastern philosophy comprises much of the history of philosophy and deserves academic attention, our aim in this book is to introduce you to some of the central concepts in the Western philosophical tradition.

Recall our definition of philosophy above, and notice that it has four components:

(i) a systematic study

(ii) of objective reality (including the tools of studying reality)

(iii) using good reasoning (the tools of logic and language)

(iv) in order to:

    (a) clarify strange and obscure questions,

    (b) solve problems that humans find significant, and

    (c) enrich human work and play.

Philosophers were among the first thinkers to study reality "systematically." A systematic study attempts to answer a series of specific questions about a subject. Systematic studies often begin by attempting to answer broad questions about the subject (e.g., Can humans know anything? Is there an external world?) and then, after establishing plausible (if not conclusive) answers, proceed to more specific questions (e.g., What can humans know from testimony? What is our external world like?).

Philosophers are primarily interested in objective reality—the way things really are. Reality includes those things and events that make up the world and claims about them. When we study reality we may ask questions about concrete concepts, such as our government, our planet, or our universe, as well as less concrete, or "abstract," ideas, like the nature of our minds, laws of nature, causation, the properties of objects, morality, or God.

To study reality systematically, philosophers rely on a set of tools to guide the process. These tools include the rules of logic and the rules of language. Logic helps philosophers determine the value of an argument or a piece of evidence. It provides guidelines for what is possible and impossible, plausible and implausible, and allows us to see more clearly the connections between ideas. The rules of language help philosophers express clearly the concepts under consideration. If a word or phrase is vague or ambiguous, the rules of language can often help us reconstruct the idea more precisely.

With these tools, philosophers can begin to address questions that are not directly available to other thinkers, such as physicists, biologists, or physicians. For instance, a physicist can define a natural law and predict how objects will act under certain circumstances. But physics cannot tell us what a natural law *is* or *why* it works or whether it could be different than it is. Clearly natural laws are not the sort of thing that can be touched or experimented on. As human beings with limited perceptual faculties, our data is always indirect; we infer the existence and strength of a natural law from the behavior of objects. But philosophers can ask the further questions about the nature of laws, how they guide causal processes, and whether a scientist's inferences about the existence and strength of laws are good inferences. Philosophers use this sort of reasoning to help answer questions that humans find significant, for instance, questions about the relationship between religion and science, the nature of morality, and the essential conditions for happiness. Clear, plausible answers to these and many other questions help enrich human experience by (a) increasing our confidence in how we approach reality, (b) helping us to recognize and be content with our limitations, and (c) helping us better understand ourselves.

As we noted, philosophers were among the first to study reality systematically. By most accounts, the first philosopher was Thales of Miletus (cir. 585 BCE, pronounced: **they**-leez). Prior to Thales' philosophical description of the world, most discussions of reality were *pragmatic* or *religious*.

Pragmatic studies of reality are not primarily concerned with the *truth* of claims, but merely their usefulness. Take, for example, *The Old Farmer's Almanac* (www.almanac.com). *The Old Farmer's Almanac* is an annually published periodical that correlates observable astronomical information with weather forecasts, ocean tides, and agriculture (when to plant, when to harvest, etc.). The *Almanac* tells us that the phases of the moon favor planting different seeds on different dates and that some seeds should be planted at night, by moonlight, while others yield better results if planted in daylight. It is irrelevant to the writers of the *Almanac* whether there is any actual connection between the phases of the moon and seed planting. They certainly don't take advantage of the standard tools (like physics or advanced meteorology) to demonstrate any such connection (though there is supposedly a secret formula for calculating the *Almanac's* predictions). All that matters is whether there are at least some practical correlations between planting and the time of day, that is, whether it is *useful* to believe there is a connection. The *Almanac's* motto highlights this approach: "Useful, with a pleasant degree of humor" (It includes no period. But you know what they mean, right?).

On the other hand, religious studies of reality *are* primarily concerned with the truth of claims, but the sources of knowledge about these truth claims are often not available to just anyone. Whereas we are told to believe a pragmatic claim because of its practical results, we are told to believe a religious claim because it has been revealed to someone by a supernatural source. Sometimes religious and pragmatic claims walk hand in hand. For instance, it was useful to the ancient Greeks to believe in the pantheon of gods, both for explaining nature (the sun moves across the sky because Helios was entrusted with the job) and for keeping society stable (play nice or Neptune might sink your boat). However, some religious claims go beyond the pragmatic to tell us about the immaterial, or spiritual, nature of humanity and prospects for life after our time on Earth. For instance, a primary Hindu scripture, the *Song of the Lord*, or *Bhagavadgita*, suggests that humans are essentially spiritual in nature, and therefore eternal. Humans are a small, but integral part of the scene of the God's creative activity, which is our universe. If people accept their role as part of the God's plan and discern one of the three paths the deity prescribes, they will be freed from the entanglements of karma in their future eternal lives. This worldview includes claims that extend far beyond the nonspiritual, finite lives we are familiar with, though the claims are not merely pragmatic; they are about reality, and are therefore, either true or false.

In contrast to both pragmatic and religious studies of reality, philosophers are concerned with the truth of claims about reality and they appeal to sources of evidence available to anyone who cares to investigate. These sources include, centrally, *reason*, *language*, and *experience*. Because these sources have a broad influence on humans, philosophy is a broad discipline, encompassing fields as varied as physics, mathematics, biology, politics, law, and morality. As philosophers investigated these fields and our understanding of them developed, these subjects became what we now call "sciences." In fact, until relatively recently, most scientists considered themselves philosophers.

The word "science," from the Latin *"scientia"* (skee-**en**-tee-a), simply means "knowledge." Galileo's writings on motion (1590) and mechanics (1600) are derived largely from the philosophical ideas of Aristotle, Archimedes, Hipparchus, and Philopponus. Similarly, Francis Bacon's development of the scientific method is explained in his philosophical work, *Novum Organum*, or *New Instrument* (1620). And, perhaps most famously, Isaac Newton's revolutionary theory of physics is called, *Philosophiae Naturalis Principia Mathematica*, or *Mathematical Principles of Natural Philosophy* (1687). Fields we now call "sciences"—physics, politics, biology, psychology—originated with philosophers. A famous twentieth-century philosopher, Bertrand Russell (1872–1970), noted that, though philosophers are often derided as asking questions without answers, this uncertainty is "more apparent than real….[A]s soon as any definite knowledge concerning any subject becomes possible, this subject ceases to be called philosophy, and becomes a separate science" (1912: 116).

It is important to understand that philosophy and science are not fundamentally distinct enterprises, but mutually informing approaches to the same types of questions. There are, of course, questions to which particular sciences are better suited (e.g., What is the charge of an electron? How do mammals reproduce?), and there are questions to which philosophy is better suited (e.g., Is the idea of an infinite set coherent? Is it morally permissible to kill someone in self-defense?). But the overlap among the methodologies and subject matter of philosophy and science is large and significant.

It is also important to understand that not every thinker concerned with the true nature of reality has been a philosopher. Ancient medicine, law, history, religion, and some mathematics began independently of philosophy. Medicine began as a *descriptive* enterprise. Ancient physicians would document the progress of a disease until a patient died or recovered, and would then use this information to name and describe illnesses (this is what you have—diagnosis—and this is how long it will take you to die—prognosis). Medicine became a *prescriptive* enterprise (hey, here's something that might make you feel better) when practitioners began using logic and language to experiment on the causes and cures of particular diseases.

Similarly, legal claims, prior to philosophy, were often evaluated by politicians designated for the purpose. Arguments may have been permitted in courts of law, but decisions were wholly at the whim of the judge. For heinous crimes, some ancient cultures developed the criterion of requiring witnesses, while others called on the gods to judge these cases, using amulets or dice to discern the gods' will. When defendants or lawyers *were* given an audience before authorities, their style of argument was often rhetorical, aimed to placate or please the judges instead of rationally arguing the merits of their case. In Plato's dialogues, the philosopher Socrates often ridicules or humiliates these "*sophists*" for their inconsistent or vacuous way of presenting their cases. This sort of elitism is probably more social than substantive, nevertheless and unfortunately, philosophers are still often perceived as elitist or snobbish.

Finally, a distinction to keep clear between a practical philosophy and a theoretical philosophy is that a *practical philosophy* is what you might write to explain how you teach (a teaching philosophy), how you do business (a business philosophy), or how you view the world (a personal philosophy). A *theoretical philosophy*, on the other hand, is a framework for investigating some aspect or feature of reality. So, a philosopher of education may have a practical teaching philosophy, but this would derive from his theoretical understanding of the nature of education. Similarly, a philosopher who owns a business will likely have a practical philosophy of business, but that would be a product of his theoretical understanding of the nature of business. Practical and theoretical philosophies are related in that the former tends to be a function of the latter, but they are not identical. In this book, we are concerned solely with theoretical philosophies.

Domains of Philosophy

There are four primary domains of philosophy: *Logic, Epistemology, Metaphysics,* and *Ethics.* A *domain* is a field of inquiry that has a special vocabulary and subject matter. From these four primary domains, dozens of secondary domains are derived: political philosophy, philosophy of religion, philosophy of physics, philosophy of mind, philosophy of law, philosophy of psychology, and the list goes on. If there is a systematic study of anything, there is probably also a *philosophy of* that study. In this book, we will discuss a few of the most significant questions about reality addressed in each of these domains. In addition to these, we devote a portion of the book to political philosophy. Many of the earliest philosophers considered the question of humans' relationship to one another and to their government of utmost importance. They brought resources from all of the four primary domains of philosophy to bear on these questions. Because of the emphasis on political philosophy among the ancients, we address some of the central questions in this area in Part V.

Logic

Logic is the branch of philosophy concerned with the principles of rational inference, and *logicians* are philosophers who study logic. The term "logic" is derived from the Greek word **λόνος**, "*logos*" (pronounced either: **low-gōs**, or **lah-goss**), which literally means "word," but has the connotation "system" or "study," implying something substantive, as in "let me give you a *word* of advice."

Logicians study the structure of claims and how this structure can be combined with logical operators (such as: and; or; if…, then…; not; if and only if) to allow us to infer new claims. This is the called the *syntax* of a language. Logicians also study the extent to which these inferences give us reasons for thinking these new claims are true, provided that our original claims are true. This is called the *semantics* of a language.

With these important tools and some rules about evidence, we can reason about any claim we come across. Logic is often thought to be the foundation on which all philosophy is laid, because it provides a strict standard against which to evaluate our inferences in any field. In Chapter 2, we will discuss some basic logical tools we will need in all the remaining chapters.

Epistemology

Epistemology is the branch of philosophy concerned with the *nature* and *extent* of human knowledge, and *epistemologists* are philosophers who study epistemology. The term "epistemology" is derived from a combination of two Greek words: **ἐπιστήμη**, "*episteme*" (pronounced: ĕpis-**tee**-may), a noun meaning "knowledge," and *logos*. The combination means: "the study of knowledge."

Epistemologists attempt to answer questions such as: "What is knowledge?" "Can humans know anything and, if so, what?" "What does it mean to be rationally justified in believing a claim?" "What counts as evidence for a claim, and when is evidence sufficient for believing a claim?" As we will see, there are serious objections to the idea that humans can know anything at all. Some argue that these "skeptical" objections are not very worrisome, offering arguments attempting to establish conclusive knowledge. Others find skeptical arguments convincing and choose, instead, to focus on less controversial questions of justification and evidence. In Chapters 3 to 5, we will discuss classical and contemporary answers to the central questions about the nature and limits of human knowledge and justification. With an informed sense of the limits of human knowledge, we can go on to ask philosophical questions about reality beyond our perceptions, for instance, about the nature of objects and morality.

Metaphysics

Metaphysics is the branch of philosophy concerned with the nature of reality, and *metaphysicians* are philosophers who study metaphysics. The term "metaphysics" is derived from the Greek preposition "*meta*," meaning "above" or "beyond," and the noun "*phusis*," meaning "nature" or "natural essence."

Metaphysicians investigate aspects of reality that aren't accessible to physicists, such as what it means for an object to have a property like "being colored orange" or "being round"; whether humans are immaterial minds or physical brains, or some combination of the two; whether an object can lose any of its parts yet remain the same object; whether humans are the sort of beings that can make unconstrained (free) choices; and whether a being like God exists.

The diversity of topics and approaches in metaphysics is dizzying. In Chapters 6 to 10, we will attempt to allay some of this confusion by distilling some of the central and most influential arguments in this broad field. We will begin with the nature of properties and how objects have them. Then we will move up the chain of "being" to humans, particularly arguments about the nature of minds, persistence over time, and free will. Finally, the section culminates with arguments about the nature and existence of the "ultimate being," God.

Ethics

Ethics is the branch of philosophy concerned with the nature of moral reality, and *ethicists*, or *moral philosophers*, are philosophers who study ethics. "Ethics" is derived from the Greek "*ethikos*" which means "concerned with character," and has its root in the word "*ethos*," which means "custom" or "habit." This root is significant in the history of philosophy because the philosopher Aristotle, in the earliest systematic investigation of morality (*Nicomachean Ethics*), argues that we build our moral character through habit, that is, by doing good or bad things. Just as we build our muscles through regular exercise, we build virtuous characters by regular virtuous actions.

By "moral reality," we mean those features of reality that constitute a demand on the actions of a certain sort of creature. Rather than simply being the way things *are*, moral reality is a set of structures that indicates how things *should be*, specifically how a certain type of being should *behave*. Therefore, whereas most claims are descriptive—claims about the way things are—moral claims are *normative*—claims about the way things ought to be.

Ethicists attempt to answer a variety of questions about morality, including questions about the nature of human value and normativity; whether there are any objective moral obligations; how humans could know the truth of a moral claim; and which actions are permissible, obligatory, or prohibited for beings like us. In Chapters 11 to 15, we will introduce you to four subfields of ethics and some of the most influential arguments in each, beginning with approaches to moral theory, proceeding to environmental ethics and animal welfare, bioethics, and, finally, political philosophy.

Political Philosophy

Political philosophy investigates the social relationships among humans, specifically how we govern ourselves in groups. Political philosophers seek answers to a number of descriptive and normative claims, including: What is the nature of humanity? Are humans essentially individual or social creatures? Are they primarily generous and altruistic or greedy and selfish? Given human nature, how should government be structured? How should humans regard government?

Political philosophers who argue that humans are essentially social creatures tend to argue that they must be governed very strictly in order to achieve their full potential. These philosophers also tend to argue that humans should view government as a parent-figure that has its "children's" best interests at heart. Citizens cannot opt out of a government any more than they can opt out of a family.

Philosophers who argue that humans are essentially individuals tend to argue that they must be governed very loosely, so that they can freely investigate and pursue what they perceive to be morally good. These philosophers offer rational and empirical evidence that governments who do not take the individual seriously face rebellion and civil war. These philosophers also tend to argue that humans should view government as a distinct individual with which they enter into a contract for certain protections and goods. At any time, humans can opt out of this contract, either by dissolving the state or by moving to a different one. There are a wide variety of positions within these broad characterizations, but these are the sorts of questions and answers that political philosophers investigate.

The Beginnings of Philosophy

The significant historical change wrought by the development of philosophy was not a change in the *content* of discourse. The primary subjects of interest among philosophers and nonphilosophers were largely the same: politics, religion, morality, the natural world, and the nature of humanity. The significant change was in the *approach* to this content, primarily away from religious and pragmatic approaches and toward a more reasoned, objective investigation.

As we noted earlier, it is widely accepted that the first philosopher was Thales of Miletus. From what we can tell, Thales argued that, though the world is made up of a diversity of materials, the ultimate source of everything—in Greek, the *arche*, or "first principle"—is water. This claim involves several interesting assumptions, not least of which is that almost everything has an ultimate source; and an interesting conclusion is that water seems to be the unique thing that does not have an ultimate source.

Little is actually known of Thales' views, and we may detect religious influences even in his claim about the ultimate source of reality, since ancient Hebrew, Egyptian, and Babylonian worldviews include claims about water that imply either that it is alive or has life-giving properties. Nevertheless, Thales' approach marks a unique change in approach to thinking about reality:

(i)   it is nonreligious (Thales made no appeal to gods, divine inspiration, or religious texts to defend his view);

(ii) it is not essentially pragmatic (the claim that the source of everything is water is irrelevant to how farmers plant and rulers rule); and

(iii) it is principled (Thales reasoned according to certain principles that are open to question and test (e.g., that everything save one has an ultimate source).

One of the best attested stories of Thales' natural acumen is that he predicted the solar eclipse of 585 BCE. Though lunar eclipses were easily predictable, solar eclipses were rarer and thought to be random. It is likely that during his travels in Egypt, Thales learned something about astronomic events that he used in Greece to predict the eclipse. Even if this is true, his use of logic to predict an astronomical event distinguishes Thales from his contemporaries.

In a related story, Plato recounts Thales' penchant for watching the stars:

> *A witty and attractive Thracian servant-girl is said to have mocked Thales for falling into a well while he was observing the stars and gazing upwards, declaring that he was eager to know the things in the sky, but that what was behind him and just by his feet escaped his notice (Plato, Theatetus, 174a, trans. Kirk and Raven, 1957).*

The most interesting stories about Thales are probably more legend than history. When ridiculed for his poverty, it is said that Thales predicted a large olive crop for the following season, raised enough money to put deposits on olive presses, hired them out, and made a large sum of money. He then gave away all his earnings to show that, though philosophers could use reason to become wealthy, they do not care for such things (Aristotle, *Politics*). Whatever the truth of the matter, stories like these served to establish, to much of history's satisfaction, Thales' title as the first philosopher.

Another early philosopher, Xenophanes (pronounced: zen-**off**-uh-neez), argued that the ancient mythical accounts of the gods are simply *anthropomorphisms* (reflections of human ideas and actions), implying more about the humans who told these stories than any supernatural reality:

> *If cattle or horses or lions had hands and could draw,*
> *And could sculpt like men, then the horses would draw their gods*
> *Like horses, and cattle like cattle; and each they would shape*
> *Bodies of gods in the likeness, each kind, of their own*
> *(from Clement of Alexandria,* Miscellanies, *5.110).*

Here, Xenophanes draws an analogy between descriptions of the gods and humans and concludes that the similarity is too striking to take seriously the idea that these descriptions tell us anything about the divine. This sort of reasoning helped establish the division between philosophical thinking and religious thinking.

However, perhaps surprisingly, it also helped shape Medieval Christian views of God. This quotation from Xenophanes was handed down through history by Christian theologian Clement of Alexandria (c. 150–c. 215 CE), who was famous for teaching that Greek philosophy was actually a preparation for the message of Jesus Christ. Clement argued that, with the arrival of Jesus, philosophers could now push beyond what they knew through reason alone and embrace the revelations of Christ. But Clement took philosophy very seriously as a limit on what humans can legitimately say about God, using passages, such as the one from Xenophanes, to guard against overconfidence in our claims about God.

It is helpful to note that, like Clement, Xenophanes was not opposed to religious claims. Clement also reports that Xenophanes said, "God is one, greatest among gods and men, not at all like mortals in body or thought" (*Miscellanies*, 5.109). Similarly, the Greek philosopher Simplicus records Xenophanes explaining of God, "But without effort he shakes all things by the thought of his mind" (*Commentary*, 23.19). So, again, the distinction between philosophical and religious thought is not its content—for many philosophers write on the topic of religion—the distinction is determined by the *approach* to that content. Philosophers rely less on subjective religious experience and folklore to inform their religious beliefs and more on reason and tradition.

A third ancient philosopher who epitomizes the emergence of this new approach to reality was Zeno of Elea (c. 490 BCE). Zeno was like the kid in class who is always trying to stump the teacher—always pushing, always challenging, and never offering any constructive suggestions. The only difference is that Zeno was incredibly good at it. Zeno challenged many widely held beliefs about mathematics and physics by constructing *paradoxes* that seem to imply that these beliefs are absurd.

A *paradox* is an apparent absurdity or contradiction that results from assuming that at least two very plausible claims are true. Take, for example, the classical *sorites* paradox (pronounced: so-**right**-eez). Consider the following two claims:

1. One grain of sand does not equal a heap of sand.
2. Therefore, adding or removing one grain of sand to or from any amount does not determine whether that amount is a heap.

Both these claims seem very obviously true. But the following claim also seems true:

3. By gathering one grain of sand at a time, you will eventually have a heap of sand.

Yet, if 1 and 2 are true, it seems that 3 could not be. This paradox is attributed to a Megarian (Euclid's school of philosophy) logician from Elea, named Eubulides.

Zeno is famous for constructing equally perplexing paradoxes, all handed down by Aristotle, for concepts associated with motion and number. In a particularly famous example, Zeno argues that, if the famous runner Achilles were to race a tortoise, and if Achilles were to give the tortoise the slightest head start, Achilles would never win:

> ...*in a race the quickest runner can never overtake the slowest, since the pursuer must first reach the point whence the pursuit started, so that the slower must always hold a lead* (Aristotle, Physics, 239b, 14).

The idea is that, while Achilles is running to reach the tortoise's location, the tortoise is moving, however slowly, so that when Achilles arrives at the tortoise's former spot, the tortoise will have moved forward, albeit a very short distance. And, while Achilles attempts to move to this new location, the tortoise will have, again, moved forward, however small a distance. As long as the tortoise keeps moving, Achilles will have to cross the distance between his location and tortoise's last location, only to discover that he has not yet overtaken the tortoise. If the line tracing the distance between Achilles and the tortoise is infinitely divisible (as we were taught in geometry), then Achilles will never overtake the tortoise. But surely this cannot be!

What are we to do with paradoxes like these? The wrong reaction would be to conclude that there is no solution and that there are no answers to any significant questions. That a puzzle is *difficult* to solve is not evidence that it *cannot* be solved. The right reaction is to take it as a philosophical challenge to be solved by careful reasoning. The value of paradoxes lies in their ability to challenge our deeply held beliefs, regardless of where we obtained them, whether from common sense, religion, or practicality. So, if we want to continue holding these beliefs and remain rational in doing so, we must accept the challenge and attempt to solve the puzzle. As it turns out, the paradox of Achilles and the tortoise led to several developments in mathematics (albeit, 1100 years later), including theories of infinite sets and calculus, which, together, show how Achilles could overtake the tortoise in such a race, despite our intuitions about infinitely divisible lines.

Still Struggling

Perhaps you're wondering: *What's all the fuss? Why spend time on these questions?* But consider a common beliefs like, "The sky is blue," or "This plane I'm on probably won't crash." How would you check to see whether these beliefs are true? With respect to the first, you typically look at the sky and perceive it to be blue. But you perceive lots of things that are false—sticks look bent in water; long, dry roads often look wet when they're not; certain paintings make objects look three-dimensional when they're only two, etc. So, how could you check your perception in this case? Ask an expert? What sort of evidence do experts point to? These are questions philosophers find important because they matter for important beliefs.

Consider, for example, the second belief above. The belief about the safety of your flight is quite important. Your life is at stake. And yet you probably base your trust in the proper workings of the plane on the belief that "most planes make their destinations safely." But that is a belief about *past* flights. What about *this* one? Have you checked the flight log, the maintenance records, or the pilots' sobriety? Without checking these things, doesn't your evidence seem quite weak? How can you evaluate the strength of this evidence? Philosophy matters because it provides a set of tools (used by scientists, lawyers, historians, and everyone else who is interested in reality) to help answer important questions.

Philosophical Eras

The history of philosophy can be roughly divided into eight eras, as shown in Figs. 1-1 and 1-2:

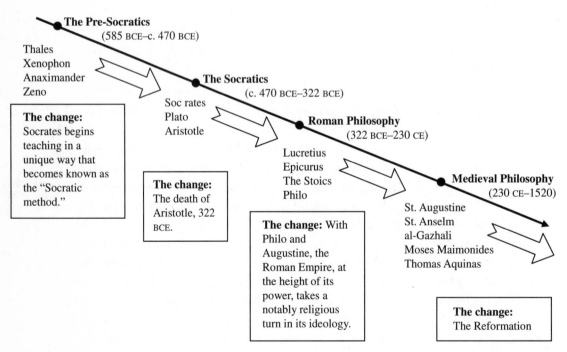

FIGURE 1-1 • Ancient and Roman philosophy.

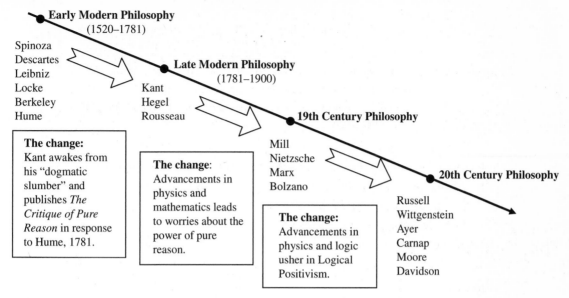

FIGURE 1-2 · Modern and contemporary philosophy.

The Value of Philosophy

Many disciplines have clear practical payoffs, e.g., engineering, business, education, and law. Philosophy's payoffs are indirect and less clear. So, why study philosophy? As with many fields in the humanities, philosophy is not vocational, it is skill-based. Writing is a craft that few have the time or energy for, and the promise of a payoff (except perhaps for novelists or journalists) is limited unless you simply enjoy doing it. The same goes for learning dead languages like Latin and Greek, and, of course, also for religion, fine arts, classics, and history. Philosophy is a skill in that it involves both clear, precise writing and a working knowledge of logic and argument strategy. The benefits of philosophy over many other fields in the humanities are its practical implications for all other fields.

Philosophy offers a powerful set of tools for forming and evaluating beliefs about reality. All thinkers use these tools so, conceivably, and with a little work, philosophers can contribute in interesting ways to any field. Twentieth-century British philosopher Bertrand Russell explains that philosophy is largely a negative enterprise, paring away our confidence in our common sense view of the world. According to Russell, philosophy reveals that our common sense beliefs about reality are more like myths or fables:

> *The value of philosophy is, in fact, to be sought largely in its very uncertainty.*
> *The man who has no tincture of philosophy goes through life imprisoned in the*
> *prejudices derived from common sense, from the habitual beliefs of his age or his*
> *nation, and from convictions which have grown up in his mind…. As soon as we*
> *begin to philosophize, on the contrary, we find, as we saw in our opening chapters,*
> *that even the most everyday things lead to problems to which only very incomplete*
> *answers can be given (The Problems of Philosophy, 1912: 118).*

Our beliefs are stymied in our commonsense perceptions of the world, engraved by the "truths" our culture and our education instill in us. Philosophy helps us break through the conceptual fog to see these ideas for what they are, whether useful fairy tales or a genuine glimpse of reality.

With the lifting of this fog comes a unique positive consequence of philosophizing. Russell explains that philosophy suggests new avenues for exploration:

> *Philosophy, though unable to tell us with certainty what is the true answer to the*
> *doubts which it raises, is able to suggest many possibilities which enlarge our*
> *thoughts and free them from the tyranny of custom…. [I]t removes the somewhat*
> *arrogant dogmatism of those who have never travelled into the region of liberating*
> *doubt, and it keeps alive our sense of wonder by showing familiar things in an*
> *unfamiliar aspect (ibid., p. 118).*

If your parents told you the Earth was flat and, through your philosophical investigation, you learn that your parents' testimony on geology and geography is not completely reliable, you may be led to suspend judgment about whether the Earth is flat. And given this new insight into testimony, you may now better understand where to look for new evidence about the claim. For instance, you may look for experts in geology and geography and ask how they would go about answering the question; you may ask about *their* evidence.

Like Russell, the eighteenth-century German philosopher Immanuel Kant viewed the process of philosophy, both in its negative and positive roles, as inherently liberating:

> *Enlightenment is man's emergence from his self-imposed immaturity. Immaturity*
> *is the inability to use one's understanding without guidance from another. This*
> *immaturity is self-imposed when its cause lies not in lack of understanding, but in*
> *lack of resolve and courage to use it without guidance from another. Sapere Aude!*
> *"Have courage to use your own understanding!"—that is the motto of enlightenment*
> *("What is Enlightenment?" 1784: 2).*

Kant argues that the ability to reflect on your beliefs without coercion requires courage and resolve in the face of the cultural influences that would lead you to complacence. The one who finds the personal strength to overcome obstacles to enlightenment (in Kant's rational sense of the term) will discover a unique kind of freedom:

> Thus, once nature has removed the hard shell from this kernel for which she has most fondly cared, namely, the inclination to and vocation for free thinking, the kernel gradually reacts on a people's mentality (whereby they become increasingly able to act freely), and it finally even influences the principles of government, which finds that it can profit by treating men, who are now more than machines, in accord with their dignity (ibid.: 7).

Kant argues that thinking rationally, guided by philosophical principles, liberates the mind in a way that reveals the unique dignity of humanity, leading to both personal and moral growth.

However, a word of warning is in order. In spite of Russell's and Kant's optimism about the value of philosophy, there are limits to our ability to reason that cannot be taken lightly. We can only reason about the *content of claims*: the content of testimony, a text, sense experience, the output of a computer program. If there is no content or if there is content that cannot be expressed in clear English sentences—for example, a private religious experience, a gut feeling, an attitude toward a traumatic event, or a decision for which the outcomes are vastly unclear—then logic and reason cannot be called upon to answer questions about these things. That is not to say that philosophers cannot argue about the nature of religious experience, but they cannot pass judgment on a *particular* religious experience. Similarly, philosophers can discuss attitudes and trauma, and their psychological impact on, say, our characters, though they cannot evaluate a *particular* response to a traumatic event. Philosopher and mathematician Blaise Pascal (1623–1662), writing on the relationship between faith and reason, noted, insightfully:

> Two excesses: to exclude reason, to admit nothing but reason (1670: fr. 253).
>
> If we submit everything to reason our religion will be left with nothing mysterious or supernatural. If we offend the principles of reason our religion will be absurd and ridiculous (1670: fr. 173).

With a clear balance of its power and its limitations, philosophy can pave the way for important answers to significant questions. We hope you find the study of philosophy engaging, challenging, powerful, and, ultimately, satisfying.

QUIZ

1. **The tradition of Western philosophy began with**
   A. Greeks such as Plato and Aristotle.
   B. Greeks such as Thales, Anaximander, and Anaximenes.
   C. Confucius.
   D. Confucius's legacy of ideas.

2. **A primary difference between Eastern and Western approaches to philosophy is that Eastern philosophers do not as firmly distinguish their broadly religious perspectives from their philosophical investigations.**
   A. True
   B. False

3. **Philosophers are primarily interested in _____**
   A. subjective reality as only they see it.
   B. people's perceptions.
   C. obeying the laws of a society.
   D. objective reality.

4. **Tools that philosophers use include _____**
   A. rhetoric and fallacious maneuvering.
   B. misleading reasoning to win arguments.
   C. the rules of logic and the rules of language.
   D. the Bible and other religious texts.

5. **Central sources of evidence for the philosopher include**
   A. reason, language, and experience.
   B. reason and language.
   C. subjective experience.
   D. the experience of an experience.

6. **In English, the Latin *scientia* means**
   A. science.
   B. truth-seeking.
   C. the systematic study of.
   D. knowledge.

7. **A practical philosophy is a framework for investigating some aspect or feature of reality.**
   A. True
   B. False

8. _____ is concerned with the principles of rational inference.
   A. Philosophy
   B. Philosophical study
   C. Logic
   D. Semiotics

9. What area of philosophy would concern itself with what it means for an object to have a property like "being colored orange" or "being round," whether humans are immaterial minds or physical brains, whether an object can lose any of its parts yet remain the same object, and whether a being like God exists?
   A. Logic
   B. Semiotics
   C. Metaphysics
   D. Epistemology

10. Who said: "Have courage to use your own understanding!"?
    A. Russell
    B. Mill
    C. Kant
    D. Pascal

chapter 2

Thinking Critically about Reality

How can we know what we should believe? What sort of standards exist by which we can test our beliefs? Philosophers discovered and developed the tools of reasoning known as logic that guide our evaluation of evidence in all areas of inquiry, from the most abstract thinking about God and mathematics to the concrete claims of science and engineering. In this chapter, we will introduce you to the most basic standards of good reasoning. These tools will guide and inform our discussions in all the remaining chapters.

CHAPTER OBJECTIVES

In this chapter, you'll learn about...

- Basic aspects of critical thinking
- Claims
- Different kinds of arguments
- Various forms of evidence
- Argument strategies

Claims

The most fundamental concept in philosophy is the *claim*. All of us make claims and are presented with claims dozens of times a day: "My car won't start," "Your girlfriend is cheating on you," "Global warming is true," "The theory of evolution is false," "Ford: America's Best-Selling Pickup Truck." A claim is *a declarative statement about reality*. It is not a question, command, or exclamation, but a statement that expresses a state of affairs, that is, the way things are. For instance, one state of affairs is that there are four *e*'s in the word Tennessee. A claim expressing that state of affairs is the English sentence, "There are four *e*'s in the word Tennessee." This claim can be expressed in a number of different languages while still expressing the same state of affairs. Since there are in fact four *e*'s in the word Tennessee, this claim is *true*, it expresses an *actual* state of affairs. Claims can also be false. For instance, the claim, "There are five *e*'s in the word Tennessee," is still a claim, and would express the same state of affairs in a number of different languages. But since it does not express an actual state of affairs, it is *false*.

Here are four examples of (what we generally take to be) *true claims*:

1. The Earth orbits the sun.

2. Yellow is a color.

3. 2 + 2 = 4.

4. Gasoline is flammable.

Here are four examples of (what we generally take to be) *false claims*:

1. The Earth is flat.

2. Yellow is not a primary color.

3. 7 − 5 = 3.

4. Gasoline is not made from petroleum.

Here are four examples of statements that are *not claims*:

1. Is the Earth tilted on its axis?

2. Do not use yellow on that painting.

3. Aha!

4. Do not drink the gasoline!

Claims can be *simple*, expressing only a single state of affairs, for example, "The Earth is flat," "The cat is on the mat," and "You are nuts." Simple claims have no "operators." *Operators* are logical devices that allow us to express relationships among claims or states of affairs. There are five standard operators: and; or; not; if…, then…; and if and only if. Further operators include *quantifiers*, such as: all; some; most; none; *modal operators*: possibly; necessarily; impossibly; and *indexicals*: then, now, when, before, after, his, hers, here, there, etc.

If a claim includes an operator, it is called a *complex claim*. For example, "The Earth is flat **and** the cat is on the mat," "Either the Earth is flat **or** the cat is on the mat," "You are **not** nuts." To speak precisely about complex claims, philosophers have named the various parts of complex claims. Studying the chart below (Fig. 2-1) will help you better understand some of the more difficult material in this book.

Arguments and Evidence

By definition, claims are about reality, and are therefore either true or false. So, which claims should we believe? "Arguments" help us evaluate the truth of claims. Arguments, as philosophers use them, are not heated debates or angry discussions. An *argument* is one or more claims (called premises) intended to support the truth of another claim (called a conclusion). The idea is to take claims we already believe (the premises) and organize them in such a way that they lend support or plausibility to another claim (the conclusion).

You may rightly ask, "If an argument requires that we begin with some claims as premises, how can we ever use arguments to support the claims we believe?" And similarly, "If we cannot evaluate the starting claims, how can we rely on claims derived from them?" The answer to the first question is clearly that we cannot support *all* the claims we believe using arguments. Arguments can only help us evaluate some claims; there must be some claims we take as "basic," evaluated in some additional way. We will discuss one of these additional ways in this section. But it is important to see that we must begin any investigation with a few claims that we presuppose, or believe from the start. This does not mean that we cannot go back and reevaluate these basic beliefs using new arguments at a later time. It just means that there are no assumptionless starting points. But, of course, the better our basic beliefs, the more faith we can rest on the claims derived from them using arguments.

Imagine we want to support the truth of the claim, "Cats are warm-blooded." This is our conclusion. One way to support the truth of this conclusion is by appealing to certain definitions, in this case, the definitions of "cat" and "warm-blooded." For example, someone could offer the following argument:

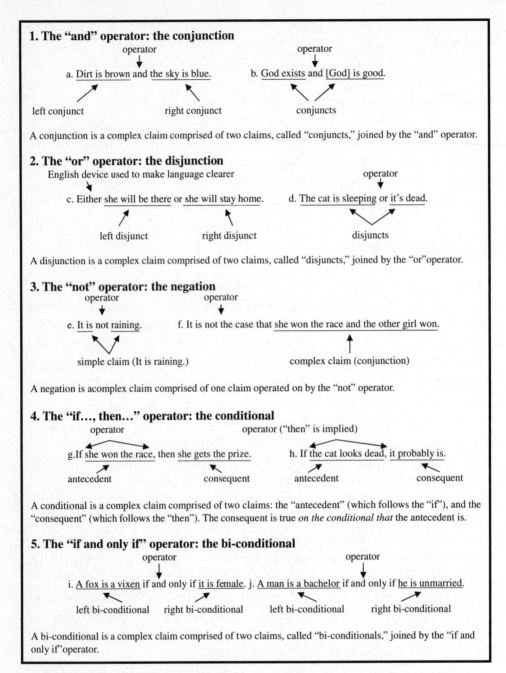

FIGURE 2-1 · Complex claims and their parts.

Argument A:

1. All cats are mammals.
2. All mammals are warm-blooded.
3. Therefore, all cats are warm-blooded.

In this argument, the conclusion, "all cats are warm-blooded," is supported by two premises: "all cats are mammals" and "all mammals are warm-blooded." We number each claim so we can refer to them later by number. This makes things easier as arguments get more complicated. The horizontal line helps distinguish premises from conclusions, and is called the *derivation line*.

There are two general types of arguments: inductive arguments and deductive arguments. The difference between them has to do with *the relationship between the premises and the conclusion*. Consider Argument A, again. It should be clear that this conclusion is fully supported by these premises—as long as these premises are true, the conclusion must be true. The conclusion follows necessarily from the premises. An argument where the premises and conclusion are related in this way is called a *valid argument*. All valid arguments are *deductive* arguments.

TERMS: Validity

The term "valid" has come to mean three things in the English language. In our everyday speech, we tend to say a claim or argument is "valid" when we mean it is legitimate or reasonable or worth considering. For example, someone might say, "You make a valid point," or "That's a valid argument."

Social scientists use the term "valid" to mean a scientific instrument that measures what it claims to measure. For instance, a thermometer that measures decibels instead of heat is invalid. But an IQ test that really measures IQ is valid.

But philosophers use the term "valid" to mean something much more specific. In logic, "valid" refers only to arguments and means, "if the premises are true, the conclusion cannot be false." In a valid argument, the conclusion follows necessarily from the premises. As you will see below, an argument can be "valid" and not worth considering or unreasonable. This is because validity is not the only feature that determines whether an argument is good or bad.

Here are two examples:

A <u>valid</u> argument:	**An <u>invalid</u> argument:**
1. All frigs are fracks.	1. If it is raining, the sidewalks are wet.
<u>2. All fracks are fredas.</u>	<u>2. The sidewalks are wet.</u>
3. Therefore, all frigs are fredas.	3. Therefore, it is raining.
In this argument, even though the premises are meaningless, if they were true, the conclusion would be true.	In this argument, even though it may sound good, the sidewalks could be wet for a number of reasons (sprinklers, garden hose, etc.). Therefore, the conclusion could be false even if the premises were true.

Consider another example:

Argument B:

1. Most Republicans are conservative.
2. <u>Lindsey Graham is a Republican.</u>
3. Therefore, Lindsey Graham is probably conservative.

In this argument, the conclusion is strongly supported by the premises, but not fully. It is still possible, even though most Republicans are conservative, that Lindsey Graham is not. He may be one of the few who are more liberal. Since you cannot be sure which, given these premises, and even if both are true, you can only conclude that Graham is *probably* conservative. The conclusion follows from the premises with some degree of probability. An argument where the premises and the conclusion are related in this way is called an *inductive argument*. All inductive arguments are *invalid*—that is, their conclusions do not follow necessarily from their premises. Nevertheless, if the premises grant a high degree of probability to the conclusion (often, more than 50% likelihood), the relationship between them is called *strong*.

The relationship between the premises and the conclusion is not the only important feature of an argument. We also need to know whether the premises are true. Consider two slight variations of Arguments A and B:

Argument A*:

1. All cats are Vogons.
2. All Vogons are warm-blooded.
3. Therefore, all cats are warm-blooded.

Argument B*:

1. Most Republicans are liberals.
2. Lindsey Graham is a Republican.
3. Therefore, Lindsey Graham is probably a liberal.

Arguments A* and B* have the exact same relationship between their premises and conclusions as their counterparts, Arguments A and B. But we would not want to put our trust in the conclusions of these arguments, given their premises. Despite the fact that A*'s conclusion is true, both of its premises are false, and therefore, do not justify belief in the conclusion. Argument B*'s conclusion is false, but the premises strongly support it. Nevertheless, premise 1 is false, so the premises do not justify belief in the conclusion.

Therefore, for an argument to be a *good* argument, it must meet two conditions: (a) the conclusion must *follow* from the premises (that is, the relationship between its premises and conclusion must be either *valid* or *strong*), and (b) its premises must be *true*. Think of these two features as terminals on a battery (see Fig. 2-2).

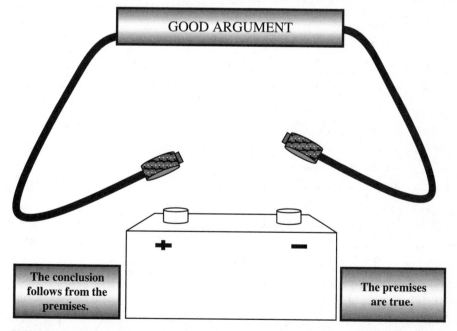

FIGURE 2-2 · A good argument requires that the conclusion follows from the premises and that the premises are true. [Figure from Jamie Carlin Watson and Robert Arp, *Critical Thinking: An Introduction to Reasoning Well* (London: Continuum, 2011).]

Both terminals must be connected in order for an argument to be good. If either terminal is disconnected, the conclusion is not justified for you by those premises.

You may know the truth-value of the conclusion on independent grounds, but this has nothing to do with evaluating the argument in front of you. The conclusion's truth-value is irrelevant to evaluating an argument for it. The premises of an argument constitute *evidence* for the truth of a claim. This evidence is your reason for believing the conclusion to be true. So, if there is no argument for a conclusion in which both conditions are met, then either you shouldn't believe the claim (it is unjustified for you) or you must have some non-argument-based evidence for the claim.

What is evidence and what counts as evidence? *Evidence* is a psychological state in which (a) a claim seems true on the assumption that a claim or set of claims is true, or (b) a claim seems true given that someone is in a particular psychological state. A claim or psychological state can be evidence even if you do not end up believing the claim for which it is evidence. For example, together, the two claims:

1. If God is all-good, he would prevent any evil he could.

2. God is all-good.

constitute evidence for the claim:

3. No evil exists that God could have prevented.

If we were convinced that 1 and 2 were true, we should accept 3. But we need not be convinced that 1 and 2 are true. Perhaps you accept 1, but have good reasons for thinking that God does not exist, so that you believe 2 is false. Or perhaps you have reasons for thinking God exists and is all-good, but that his goodness would lead him to allow some evil in order to accomplish greater goods, so that you believe 1 is false. Nevertheless, you can still see that 1 and 2 constitute evidence for 3. You may not find that evidence *sufficient* for believing 3, given that you reject one of the premises, but they are evidence, nonetheless.

Similarly, the psychological state of "seeing the coffee-shop wall as brown" constitutes evidence that the wall *is* brown. Under normal circumstances, this evidence might be sufficient to justify your belief that it is brown. But imagine that the manager, whom you know and trust, tells you that the wall was recently painted red and a purple bulb nearby still makes it look brown. Now you have reasons for doubting your initial evidence. You may now believe the wall is red. Nevertheless, the wall still seems brown, and that seeming is still evidence that it is brown, even if you no longer consider that evidence sufficient for believing it is brown.

We can identify different types of evidence by distinguishing four categories under which they can be organized: direct and indirect, experiential and non-experiential. These distinctions can be a bit tricky, but they will be important in the following chapters.

Direct evidence is evidence that it immediately available to you, for instance, seeing an object as red and round, smelling it as an apple, feeling it as smooth and waxy, etc. Sensory evidence is commonly categorized as direct evidence. Rational intuitions are also considered direct. "Rational intuition" does not refer to the vague, flight of fancy often associated with the paranormal or "women's intuition," but a state of its seeming to a person that a claim is true and could not be false, that is, that the claim is necessarily true. Most people have this sort of psychological state about the claims, "2 + 2 = 4," "all bachelors are unmarried," and "no round squares exist." These intuitions are often the primary source of evidence in mathematical, logical, and ethical reasoning.

Scientific experiments are a primary example of indirect evidence. No one sees ibuprofen reducing swelling or reducing pain, but there is indirect evidence that it does, given that we experience a decrease in swelling or pain soon after taking ibuprofen. All scientific experiments work this way and all are indirect. Memory is another indirect source of evidence. Memory is directly available, just like the *results* of a scientific experiment, and the results are about something in the past, also like a scientific experiment. Since, in both cases, the results are one step removed from the actual event to which they testify, both are indirect sources of evidence. Testimonial evidence is also indirect, for the same reasons.

The distinction between experiential and nonexperiential evidence is more controversial. Some philosophers argue that all evidence is experiential (derived from our experience of the world). These philosophers are called "empiricists." Sense experience and scientific experiments are the least controversial sources of experiential evidence. Other philosophers argue that at least some evidence is not derived from our experience of the world, that is, it is nonexperiential. These philosophers are called "rationalists." The primary source of nonexperiential evidence for rationalists is rational intuition. But some rationalists, such as Descartes, also consider introspection to be nonexperiential. And some contemporary rationalists argue that memory and testimony are nonexperiential. We highlight these distinctions in the chart below (Fig. 2-3), placing "according to some philosophers" next to the more controversial categorizations. The distinction between experiential and nonexperiential evidence will become clearer in our discussions of Descartes' and Locke's responses to skepticism.

```
I. Experiential
        A. Direct Experiential Evidence
                i. Sense experience
                ii. Introspection (according to some philosophers)

        B. Indirect Experiential Evidence
                i. Scientific Experiment
                ii. Memory (according to some philosophers)
                iii. Testimony (according to some philosophers)

II. Nonexperiential
        A. Direct Nonexperiential Evidence
                i. Rational Intuition
                ii. Introspection (according to some philosophers)

        B. Indirect Nonexperiential Evidence
                i. Memory (according to some philosophers)
                ii. Testimony (according to some philosophers)
                iii. Mathematical Proofs (according to some philosophers)
```

FIGURE 2-3 · Types of evidence.

 Still Struggling

In the midst of all these details it is easy to miss the big picture. In philosophy, arguing is about getting an accurate picture of reality, whether you are defending your own claim or criticizing someone else's. And you argue every day, for instance, when you're trying to convince your roommate that he broke his promise, to convince a teller the bank made an error in your account, to convince your boss that your money-saving idea will work, or to convince a friend that some political policy will not be as effective as a politician claims. You argue every day, but do you argue *well*? The details about claims and arguments explained in this chapter are aimed to help you argue more effectively.

Argument Strategies

The most important aspect of philosophy is convincing yourself and others that a claim is true or false. You may not end up with certainty, but hopefully you will end up with the best reasons available for believing or not believing a claim.

An argument strategy is the process of using an argument to convince a particular audience—yourself or others—that a claim is true or false.

There are a variety of ways to use arguments to support a claim. Some argument strategies are *direct*, which means that claims are organized in a way that provides direct support for another claim. These arguments are then evaluated in terms of whether the conclusion follows from the premises in the right way (either validly or strongly) and whether those premises are true. Other argument strategies are example-based, which means that an argument for a claim is evaluated in terms of a parallel argument or case study to increase or decrease the plausibility of the original argument. This second argument strategy is called a *thought experiment*. Still other strategies are comparative, which means the strengths and weaknesses of one argument are evaluated in light of the strengths and weaknesses of another; as an example, they may be evaluated according to how well they express certain rational virtues, for instance, they require fewer things to exist, or one explains more than the other. This third strategy is known as *inference to the best explanation*. There are additional strategies, but these are three of the most common.

Except for deductive versions of the direct argument, these strategies are all variations of inductive argument, which means that all conclusions will be more or less probable, and not certain. Nevertheless, these strategies have proven themselves in the most successful and crucial contexts of human existence, including medicine, law, physics, and engineering.

Direct Argument

In a direct argument strategy, philosophers construct arguments using premises their audience (hopefully) already find plausible. If an arguer can organize these accepted premises in such a way that a desired claim follows validly or strongly, the arguer has given his audience a reason to accept his conclusion. There are dozens of ways to organize an argument so that the conclusion follows validly or strongly; this organization is called the *form* of the argument. Here, we will cover six of the most common argument forms.

The first three forms are valid deductive arguments, which means, if the premises are true, the conclusion must be. The second three forms are inductive arguments, which means, if the premises are true, the conclusion follows with some degree of probability. If the probability is high enough, the argument is strong.

The first valid argument form that will be useful in this book is called *modus ponens*, which is Latin for, "mode of putting":

1. If p, then q.

2. p.

3. Therefore, q.

p and q are variables, and any claim (simple or complex) can be inserted in their places. As long as the claim you insert for p in premise 1 is the same claim you insert for p in premise 2, and as long as you do the same with q in premise 1 and the conclusion, then the conclusion follows from the premises with necessity. If you learn the antecedent of premise 1 (p), which is premise 2, then from premises 1 and 2, you can derive the consequent of premise 1 (q) as the conclusion (see Fig. 2-4). If the premises are true, then the conclusion must be true, as well.

p q 1. If it is raining, then the sidewalk is wet.	1. If it is green, then it is colored.
p 2. It is raining.	2. It is green.
q 3. Therefore, the sidewalk is wet.	3. Thus, it is colored.

FIGURE 2-4 · Two examples of modus ponens.

The second valid argument form we will refer to later in this book is called *modus tollens*, which is Latin for, "mode of taking":

1. If p, then q.

2. Not q.

3. Therefore, not p.

Like before, p and q are variables to be replaced with meaningful claims. If you learn that the consequent of premise 1, q, is denied, or "negated," then if both premises 1 and 2 are true, you can then derive the negation of premise 1's antecedent, p, as the conclusion (see Fig. 2-5). Again, if the premises are true, then the conclusion must be true, as well.

1. If it is raining, the sidewalk is wet. 2. The sidewalk is not wet. 3. Therefore, is it s not raining.	1. If it is green, then it is colored. 2. It is not colored. 3. Thus, it is not green.

FIGURE 2-5 · Two examples of modus tollens.

The third argument form that will be helpful later on is called *disjunctive syllogism*. A disjunction, as we explained in the chart above, is simply a complex claim comprised of two claims joined by the "or" operator. A syllogism is a valid argument with two premises and a conclusion. The form of a disjunctive syllogism looks like this:

1. Either *p* or *q*.	1. Either *p* or *q*.
2. Not *p*.	2. Not *q*.
3. Therefore, *q*.	3. Therefore, *p*.

A disjunctive syllogism has a disjunction as one of its premises, in this case, premise 1. If you learn the denial of either of the disjuncts, you can derive the other disjunct—that is, if you learn either not *p* or not *q*, you can conclude *q* or *p*, respectively (see Fig. 2-6). If the premises are true, the conclusion must be.

---

1. It is either a rat or a mouse.
2. It is too big to be a rat (i.e., it is not a rat).
3. Thus, it is a mouse.

1. It is either not green or it is colored.
2. It is not not green.
3. Hence, it is colored.

The double-negative in premise 2 is intentional and necessary for this argument to be valid. In premise 1, the first disjunct is a negation. To negate this negation, we must add a second "not." Logically, this construction entails: "It is green."

FIGURE 2-6 • Two examples of disjunctive syllogism.

The next three forms are inductive. The first inductive form is called *enumerative induction*, which involves deriving a claim about a future state of affairs from a series of past instances of similar states of affairs. For example, if at 1:00 you saw a mallard duck floating on a river, then you saw another at 1:15, and another at 1:25, you might draw the conclusion that *the next duck* I see will probably also be a mallard. You could not be *certain*, from just three instances, that the next duck you see will be a mallard, but the previous instances *increase the probability* that it will be (see Fig. 2-7).

---

1. After I drank milk on Wednesday, my stomach hurt.
2. After I drank milk on Friday, my stomach hurt.
3. Thus, if I drink milk today (Monday), my stomach will probably hurt.

1. Every time I touched this wire to this battery, a light came on.
2. Therefore, if I touch the wire to the battery again, a light will probably come on.

In this argument, premise 1 implies a series of past instances. To know how strong this argument is, we would need to know how many instances there were. The more, the better.

FIGURE 2-7 • Two examples of enumerative induction.

The form of an enumerative induction looks generally like this, though the number of instances and wording may be different:

1. At time t, I observed that X is A.

2. At time $t + 1$, I observed that X is A.

3. At time $t + 2$, I observed that X is A.

4. Therefore, probably at $t + 3$, I will observe that X is A.

Imagine, now, that you don't know much about ducks, and at 1:00 you see a bird floating down the river. You notice that it has a bright green head, a white stripe around its neck, and a bluish section on its wing. The person with you explains that it is a male duck called a "mallard." At 1:15, you see another bird coming down the river that has a bright green head, a white stripe around its neck, and a bluish section on its wing. You conclude, based on the similarities, that this, too, must be a mallard. In this case, you are drawing a conclusion based on the similarities between the duck you have seen and the next duck you come across. These similarities are called *analogues*, and this third argument form is called *argument from analogy*. You cannot be certain that the second duck is a mallard (perhaps you focused on the wrong features, say, webbed feet and wings, and concluded it was a mallard), but the number of similarities increase the probability that it is (see Fig. 2-8).

1. Rock A has a cylindrical handle with a blunt end and a sharp edge, and is a tool. 2. Rock B also has a cylindrical handle with a blunt end and a sharp edge. 3. Thus, Rock B is probably also a tool.	1. Seven patients have experienced a sore throat, fever, a cough, and nausea, and they all had the flu. 2. Patient number eight has a sore throat, a fever, a cough, and nausea. 3. Hence, patient eight probably also has the flu.

FIGURE 2-8 • Two examples of argument from analogy.

The general form of an argument from analogy looks as follows, though the number of features and objects compared may vary:

1. Object 1 has features w, x, y, and z.

2. Object 2 has features w, x, and y.

3. Therefore, object 2 probably also has feature y.

Finally, imagine you've seen dozens of mallards, and every one you've seen zigzags as it searches for food on top of the water. From these dozens of instances, you infer that all mallards must behave this way—that is, you generalize from the behavior of a sample of mallards to the whole population of mallards.

1. Of the 300 males we tested, 75% experienced a negative side effect of the drug.	1. Out of 150 pregnant women, 80% wanted to give birth naturally.
2. Therefore, probably 75% of all males will experience a negative side effect of the drug.	2. Therefore, probably 80% of pregnant women will want to give birth naturally.

FIGURE 2-9 · Two examples of inductive generalization.

This form of inductive argument is called *inductive generalization*. Inductive generalizations always include two groups: a population (males over 30; children in the 6th grade; mallards) and a sample of that population (200 males over 30; 40 children in the 6th grade; 70 mallards) (see Fig. 2-9). In order for a generalization to be a good generalization (and therefore, a strong inductive argument), the sample must meet four conditions: it must (a) be proportionate (the larger the better); (b) random (selected arbitrarily, so that the sample is not biased); (c) obtained using a valid instrument* (for instance, a questionnaire that asks the right sort of question); and, (d) obtained using a reliable instrument (an accurate measuring device).

The general form of an inductive generalization looks as follows:

1. Of $X\%$ of the population, $Y\%$ are Z.

2. Therefore, probably $Y\%$ of the whole population are Z.

Thought Experiment

Thought experiments are hypothetical cases constructed to either (a) show that a claim is plausible, or (b) show that a claim is implausible. A useful type of thought experiment is called a *counterexample*. A counterexample is an example used to show **either** that *a claim is false* or that *an argument is not good*. A counterexample to a *claim* is an argument showing that a mutually exclusive claim is true. Two claims are mutually exclusive if they cannot both be true at the same time, for instance, "the house is red right now" and "the house is blue right now"; "Michael Jackson had no sisters" and "Janet Jackson was Michael Jackson's sister." If there is a better argument for the latter claim than the former, since they cannot both be true, we have reason to believe the former is not true. Here's an example from ethics:

---

*This is a different use of "valid" than the one we've been using. Rather than meaning that a conclusion follows necessarily from a set of premises, here it means an instrument that measures what it claims to measure. See the box above: "Terms: Validity"

Argument 1:

1. Intentional killing is wrong.
2. Abortion is an intentional killing.
3. Therefore, abortion is wrong.

Counterexample to premise 1 of Argument 1:

4. It is not morally wrong to kill a mosquito.
5. Killing a mosquito is an intentional killing.
6. Therefore, some intentional killings are morally permissible.

This counterexample highlights the inconsistency between the claims "Intentional killing is wrong" and "some intentional killings are morally permissible"; both cannot be true. Since most people have good reasons to believe there is no moral harm in killing a mosquito, premise 1 is much too strong, and therefore, we have a reason to reject it. If an argument could be made for protecting mosquitoes from intentional killings (that is, an argument that premise 4 is false), this counterexample would not work.

TIP   *There would also be no problem if premise 1 read, "Intentional killing is often wrong," or "Intentional killing is sometimes wrong." But if the arguer defended that as a premise, the conclusion would also change. Now the conclusion would be, "Abortion is often wrong" or "Abortion is sometimes wrong." But almost everyone could agree with the latter and many, perhaps, with the former, so there would be little need for argument. The whole point is to determine the specific circumstances under which abortion is morally permissible and impermissible.*

Using the same example, consider a counterexample to an argument:

Argument 1:

1. Intentional killing is wrong.
2. Abortion is an intentional killing.
3. Therefore, abortion is wrong.

Counterexample to Argument 1:

1. Intentional killing is wrong.
7. Killing bacteria with antibiotics is an intentional killing.
8. Therefore, taking antibiotics is wrong.
9. But we have good reasons to believe it is not wrong to kill bacteria.
10. Therefore, there might be reasons for thinking intentional killing is not wrong in the case of abortion.

In this case, the counterexample shows that inserting a different action into premise 2 yields an unfortunate conclusion, namely, that we are never morally permitted to take antibiotics. But if these premises do not justify the claim that taking antibiotics is wrong, they do not justify the claim that abortion is wrong.

The problem with Argument 1 is probably the same, namely, premise 1 is false, but there are two at least two ways to show this: (i) argue directly that premise 1 is false (counterexample to a claim), or (ii) argue that substituting a different action leads to an undesirable conclusion (counterexample to an argument).

There are two general types of counterexample to an argument: The Technique of Variant Cases and The Bare Difference Argument. In *The Technique of Variant Cases*, an argument is constructed that has all the same elements as the argument under consideration *except* the act in question. So, the above counterexample to argument 1 is a variant case counterexample; we just substituted "killing bacteria" for "abortion" to show that the conclusion does not follow from the premises. If someone could offer good reasons for thinking that taking antibiotics is morally impermissible or that taking antibiotics is not a case of intentional killing, this counterexample would not work.

In *The Bare Difference Argument*, an argument is constructed that has *none* of the same elements as the argument under consideration except the act in question. Consider, one last time, Argument 1:

Argument 1:

1. Intentional killing is wrong.

2. <u>Abortion is an intentional killing.</u>

3. Therefore, abortion is wrong.

Bare difference counterexample to Argument 1:

11. Killing in self-defense is not always wrong.

12. <u>Abortion when the mother's life is threatened is killing in self-defense.</u>

13. Therefore, abortion is not always wrong.

In this case, we have changed the circumstances; now we are not simply considering killing, we are considering cases of self-defense. In light of common views about self-defense, abortion is morally permissible. Therefore, we have provided reasons for doubting the conclusion of Argument 1. If someone could show that killing in self-defense is always wrong or that we never need to perform abortions in defense of a mother (for instance, perhaps we could always do a Caesarian), then this counterexample would not work.

Inference to the Best Explanation

An explanation is, primarily, a claim or set of claims that answers the questions "Why?" or "How?" For example, why is a shadow of an object a certain length? The explanation would combine a claim about the size of the object with a claim about the angle at which light hits the object. Why is the fruit in the bowl on a table always fresh? The explanation would combine a claim about someone's intentions and someone's actions (maybe the same person; maybe not). *Explanations* (also called the "explanans") can serve as the premises of arguments for claims that need an explanation (also called the "explanandum").

In some cases, two sets of explanans sufficiently explain one explanandum, though both sets of explanans cannot be true. For example, imagine you are home alone late one night, when you hear scratching at the window. This is your explanandum. You quickly consider two possible explanations: The wind is blowing a tree limb against the window or one of your friends is trying to scare you.

1a. The wind is blowing a tree limb against the window.	**1b.** My friend is trying to scare me by scratching at the window.
2. I hear scratching at the window.	**2.** I hear scratching at the window.

Both explanations are sufficient for the conclusion; that is, if the premises were true, each would grant some probability to the conclusion. But it is unlikely that both explanans are true. If your friend gets to the window and the wind is already blowing a limb against it, it is unlikely that he would think it would scare you to add to the scratching.

Now you face two competing explanations for the same phenomena, so you must consider which of the two explanations is the best. (In this example, you could always go look, but there are many times when that option is not available and, therefore, we must make a decision based on other evidence.) This type of argument is called an *inference to the best explanation*. If one explanans is more probable than the other, then your inference to the best explanation is straightforward. For instance, if you suddenly remembered that your prankster friend is out of town, then the probability that it is the wind goes up, and your inference to the best explanation is complete. Other times, however, both explanans seem equally probable. In these cases, philosophers must rely on what are called the "theoretical virtues" of each explanans.

A *theoretical virtue* is a feature of an explanation that gives it its strength as an explanation. Theoretical virtues include: simplicity (the number of causes or mechanisms required for the explanation to work—the fewer the better); conservatism (the number of well-established beliefs that we would have to give up for the explanation to work—the fewer the better); explanatory scope (how many similar phenomena it can explain—the more the better); and fecundity (how many new research possibilities the explanation opens up—the more the better).

An example of an inference to the best explanation where competing explanans are roughly equally probable is a case where a small rock has broken a window on your car and you must decide whether it was intentional or whether it was caused by a passing lawn mower. Each explanation is sufficient for the broken window, you did have your lawn mowed that day, and there are children in the neighborhood who occasionally break windows. In reasoning about their virtues, you consider that they are both sufficiently simple, since neither requires causes that were not present. Both are fairly fecund—you can ask around the neighborhood about the type of damage the nefarious children have been causing and contrast this with the type of damage, if any, caused by the mowing crew. Both are also fairly conservative; your beliefs about mowers and nefarious children are consistent with the broken window. However, you know the children in the neighborhood, and it would be a slight stretch to believe they intentionally targeted your windows. So, the mower explanation is slightly more conservative than the intentional explanation. Additionally, the explanatory scope of the intentional rock-throwing is narrower than the mower. This is because you know that, although children can break car windows, doing so with a small rock is much more difficult for a child than for a lawn mower. Therefore, on this assessment of the virtues of the competing explanations, you would be wise to infer that it was the mower and begin looking for further evidence to support this conclusion.

Fallacies

A fallacy is an error in reasoning. A fallacy can be *formal*, in which case, a conclusion doesn't follow from a set of premises because the argument was organized poorly. Or a fallacy can be *informal*, in which case, a conclusion doesn't follow from a set of premises either because the premises are irrelevant to the conclusion or because the meaning of one or more of the terms in the premises prevents the premises from providing sufficient support for the conclusion.

We will explain six common informal fallacies that will be particularly useful in the remainder of this book. We will need to introduce a few more along the way, but these will serve as a foundation for identifying others.

Appeal to the People

You may remember a parent or guardian responding to one of your childhood pleas to join your friends by saying, "If everyone jumped off a bridge, would you?" Your parents were trying to help you avoid one of the most popular mistakes in reasoning, the *appeal to the people* fallacy. Lots of people believe false claims. Therefore, that a group of people believe a claim does not constitute evidence that the claim is true.

We now know that few cultures actually believed that the Earth is flat; however, it is easy to imagine that they did. But it is important to see that, even if they did, this is not evidence that it is true that the Earth is flat—lots of people can be wrong! The same goes for claims like: "No one believes premarital sex is immoral," "Most people think abortion is morally permissible," "Lots of people believe that Euclidean geometry is the only kind of geometry" (it's not, by the way), and "Almost everyone believes in evolution." If lots of people believe something, you may want to spend a little time trying to figure out why. You may discover good evidence along the way. But the fact that many or most people believe something does not count as good evidence.

> **Examples of the Appeal to the People Fallacy:**
>
> 1. "Ford: America's Best-Selling Pickup." [Why should you buy a Ford pickup truck? *Because* most people are buying them.]
> 2. "Reading: Everybody's Doing It!" [A real message from PBS. Hundreds of years of parenting undermined!]
> 3. "Almost every culture has believed in some creator deity. Therefore, something like God must exist."

Appeal to Snobbery/Vanity

Advertisements often attempt to convince you to believe or buy something by telling you that you will be stylish or attractive if you do, or to convince you not to believe or buy something by telling you that you will be unfashionable or unattractive if you do. If someone appeals to either of these reasons, she is committing the fallacy of *appeal to snobbery or vanity*.

As we saw in the previous fallacy, a claim's truth does not depend on what lots of people think about it. It also does not depend on what a particular group of people think about it—even if you *really* like them. A claim may not be in vogue or prestigious or attractive (for instance, that people have no right to health care, or that God does or doesn't exist, or that much of alternative medicine is junk), but you have a rational duty to follow the evidence where it leads, not to be popular or to avoid ridicule.

> ### Examples of the Appeal to Snobbery/Vanity Fallacy:
> **1.** "Superman Watches: Live a Life of Luxury and Distinction."
> **2.** "Professor Dawkins is a raging atheist. Therefore, to believe anything he says would be conceding atheism."
> **3.** "If you distrust scientists on global warming, then you are a narrow-minded, backward-thinking conservative."

Appeal to Inappropriate Authority

Most of our beliefs are justified by appeal to others, hopefully people in the know—authorities. But not all authorities are equal. Your parents are probably good authorities on social values—don't lie, cheat, or steal unless someone's life depends on it, etc.—but probably not good authorities on the latest advancements in climatology (unless, of course, they are climatologists).

It is important to keep in mind that, just because someone is an authority in one area, does not mean he is an authority in the area under discussion. Your physics professor may be a good authority on the laws of motion, but is probably not a good authority on the literary value of Ayn Rand novels. Similarly, just because someone claims to be an authority, doesn't mean she is. The *Journal of Paranormal Research* regularly publishes articles by people claiming to be experts in paranormal psychology. But paranormal psychology is not recognized by any respected scientific organization. Appealing to the claims of people who are not experts in the area relevant to a claim is a fallacy known as *appeal to inappropriate authority*.

An authority can be inappropriate in one of two ways. The authority can be biased, as in the *Journal of Paranormal Research* publishing something on the success of ESP in detective work, or Tiger Woods suggesting I use his own brand of golf clubs. Alternatively, the authority can be irrelevant, as in the *Journal of the American Medical Association* (a well-respected medical journal) publishing something on an interpretation of a New Testament text, or Tiger Woods suggesting that I buy a Tag Heuer watch. When evaluating the testimony of an "expert," make sure that person really is an expert and is an expert in the right field.

> ### Examples of the Appeal to Inappropriate Authority Fallacy:
> **1.** "The *Journal for the Promotion of the Environment* says that 25% of our carbon footprint comes from appliances being plugged in, even when they're not being used."
> **2.** "Leonardo DiCaprio says you should buy a Tag Heuer watch."
> **3.** "Oprah Winfrey says Dr. Robinson's new book is the most effective diet book on the market."

Appeal to the Person (Character or Circumstances)

Sometimes, an arguer will appeal to someone's character or circumstances in order to persuade you that some claim they make is true or false. For instance, if a suspect is being tried for murder and claims to not be guilty, a clever lawyer might try to convince you that her client is not guilty of murder because he was raised in an upper-middle-class family, with excellent parents, who constantly modeled a life of virtue. Therefore, her client would not lie about his innocence of this crime, consequently, he did not commit the crime. In this case, the lawyer is appealing to her client's circumstances to convince you to believe his claim of innocence. She might also point to his outstanding character, explain that he is a member of many civic clubs, a contributor to many charities, and well-known throughout the community as a completely honest man, and on these grounds, conclude that he could not be the killer. The problem is that, even if all these claims are true, the suspect's circumstances and character are irrelevant to whether, in this case, he committed the crime. His reasons for lying may outweigh his motivation for being honest; and his reasons for killing may outweigh his motivation for being virtuous. It is his lawyer's job to prove this isn't the case. In drawing inferences from his circumstances and character she is committing the fallacy of appeal to the person (often called "*ad hominem*," which means "to the man" in Latin).

Appeals to the person can be used to defend someone's claim or to discredit it. For instance, former U.S. President Bill Clinton cheated on his wife with an intern, and many of his political opponents argued that this character flaw undermined his ability to perform his duties as president. Similarly, when U.S.

President Barack Obama was campaigning for the office, some doubted his ability to act on behalf of the lower classes, given that he was raised in an upper-class family. Since character and circumstances are irrelevant to the truth of a claim, each of these complaints commits the fallacy of appeal to the person.

There are times when character is relevant. For instance, given the controversy surrounding his presidency, it is now widely believed that Bill Clinton is a liar. This character flaw gives us reason to suspend judgment about anything he says. We should not reject it outright—he might be telling the truth—but we should seek out independent evidence to corroborate before we believe. Therefore, character is relevant when evaluating a claim relevant to the character trait. So, honesty is relevant when evaluating whether to trust someone's testimony.

BOX 2-1 Appeals to Pity

Consider our murder trial, again. If, instead of arguing that her client is not guilty because his character would prevent him from committing the crime, she argued that he should not be held accountable for the crime because of his soft heart and frail body, so that any prison sentence, no matter how short, would be a death sentence, she would be committing a different fallacy, called an *appeal to emotion* or *pity*. His pitiable circumstances do not constitute evidence for or against his guilt, though they may be relevant for determining his punishment.

The SPCA commits the appeal to pity when they attempt to convince you to give money to stop animal abuse by showing picture after picture of sad-looking, abused animals and playing Sarah McLachlan's song "Angel." To be sure, you probably *should* do something to help stop animal abuse, and giving to the SPCA may be one way to do that. But *that you feel bad* is not a reason to believe these things. Therefore, in attempting to convince you by making you feel bad, the SPCA commits the appeal to pity fallacy.

Examples of the Appeal to the Person Fallacy:

1. "I ate dinner with Senator Roberts last night, and he was rude and obnoxious. You should not vote for someone like that."

2. "Superintendent Jones says the public schools in our area are at least as good as any private school. But she only attended public schools, so she could not possibly understand the difference."

3. "President Clinton is a philanderer and an adulterer. Surely, you cannot elect him to a second term."

Straw Man

Imagine playing chess with someone who is at least as good as you are. Now, imagine that, in the middle of the game, your opponent moves to a different chess board, in a different part of the room, and checkmates the person playing your color on that board, then declares victory over *you*. It is likely that you would find this quite strange and not at all a legitimate victory. This "win" is hollow—your opponent didn't beat you; she beat someone else! Nevertheless, clever reasoners use this very trick effectively against their opponents.

Imagine that some governor argues that allocating additional funds for education will bankrupt the state; therefore, the state's congress should not pass a bill allocating additional funds for education. And imagine that, in response, someone argues that the governor's plan to gut the educational budget of the state is deplorable and will have far-reaching negative consequences for the children of the state.

The problem with this response is that it does not address the governor's argument at all. The governor did not propose gutting the education budget; he proposed not allocating *additional funds* for the budget. Those predisposed to reject the governor's proposals might be taken in by this argument, but it is no different than your chess opponent's "beating you" on a completely different chess board; you didn't make those moves. When an arguer changes his/her opponent's argument to make it easier to refute, he/she has committed the *straw man* fallacy.

Examples of the Straw Man Fallacy:

1. "Pastor Jones says that most abortions are morally impermissible. But to make all abortions illegal would be unreasonably dogmatic and would ignore the complexity of real-life situations."

2. "President Bush labeled Islamic Fundamentalist Terrorists the 'Axis of Evil.' But to demonize a whole religion because of the bad actions of a few of its members is naïve and unfair."

3. "Democrats argue that we need a systematic withdrawal strategy from Iraq. But folks, that sort of cut-and-run strategy is unpatriotic and runs contrary to the freedom we value as a country."

Begging the Question

At some point, if you haven't already, you will ask someone how they know something, only to get the answer, "Because I know." This response is clearly not a sufficient answer to your "how do you know" question—the argument is circular, it assumes what it needs to prove:

<u>**1.** I know.</u>

2. Therefore, I know.

We call this argument *circular* because, until one of the claims is justified by independent evidence, the lines of support move both from the premise to the conclusion (because, if true, the premise entails the conclusion), and from the conclusion to the premise (because you must already know the conclusion is true in order to believe the premise).

Circular arguments commit the fallacy called *begging the question.* An argument begs the question if it assumes or implies in the premises what it is attempting to support in the conclusion. A classic example of a question-begging argument is the support derived from what is broadly known as "intuition," or "gut-feeling" (this is different from what philosophers call "rational intuition"):

<u>**1.** I have this feeling about them.</u>

2. I just know they are going to be married for a long time.

This sort of argument does not provide support for the conclusion at all. It is tantamount to saying: "I believe *X*, therefore I believe *X*."

BOX 2-2 The Phrase, "That Begs the Question."

You have probably heard a journalist or news correspondent say something like, "The senator responded so-and-so, but that begs the question, 'Is so-and-so correct?'" This use of "begging the question" is not how philosophers use the phrase.

"Begging the question" has come to have two meanings in contemporary English. The first is logical, and refers to a circular argument—assuming in the premises what you are trying to justify in the conclusion. The second is rhetorical, and refers to raising an additional question, or implying some other claim that requires evidence. Both are now accepted, but because of this variation, philosophers only use the phrase in its logical sense.

Examples of Begging the Question:

1. "Of course Allah exists. The Koran says so and Allah wrote the Koran."

2. "Animals have rights, and the most fundamental right is the right to life. Therefore, it is immoral to deprive an animal of the right to life.

3. "Dr. Robinson, a noted dietician, says that his book is the most effective diet book on the market. How can you deny it? He's a doctor in the field!"

Still Struggling

After seeing how many ways an argument can go wrong, you may be thinking: *People commit fallacies all the time! How do we deal with it?*

Keep in mind that, just because someone has committed a fallacy doesn't mean their conclusion is false, only that they have not offered sufficient evidence for their conclusion. Therefore, the most effective argument strategy is not simply to point out the fallacy and move on, but to fully investigate the matter. Respond to the fallacy, then highlight some appropriate evidence. Maybe your opponent is right, in which case, you learned something. But if your opponent's conclusion is not supported by the best available evidence, then you have shown that there is firm ground for rejecting his claim.

QUIZ

1. **The most fundamental concept in philosophy is**
    A. the argument.
    B. the evidence.
    C. the claim.
    D. the philosopher's idea.

2. **It's not the case that the following is a claim: "Do not use yellow on that painting."**
    A. True
    B. False

3. **All, some, most, and none are**
    A. operators.
    B. modal operators.
    C. quantifiers.
    D. indexicals.

4. **"It is not raining" is not a complex claim.**
    A. True
    B. False

5. **The conclusion supports the premises in an argument.**
    A. True
    B. False

6. **There are two general types of arguments:**
    A. inductive arguments and deductive arguments.
    B. inductive arguments and valid arguments.
    C. inductive arguments and strong arguments.
    D. deductive arguments and valid arguments.

7. **The following argument is valid:**
    Premise 1: Most Republicans are conservative.
    Premise 2: Lindsey Graham is a Republican.
    Conclusion: Therefore, Lindsey Graham is probably a conservative.
    A. True
    B. False

8. _____ is a psychological state in which (a) a claim seems true on the assumption that a claim or set of claims is true, or (b) a claim seems true given that someone is in a particular psychological state.

   A. Belief
   B. Perception
   C. Opinion
   D. Evidence

9. *Modus ponens* does not contain a negation.

   A. True
   B. False

10. _____ is used to show either that a claim is false or that an argument is not good.

    A. *Modus ponens*
    B. *Modus tollens*
    C. Counterexample
    D. Thought experiment

Knowledge and the Problem of Skepticism

What does it mean to "know" something? Can you honestly say you "know" that 2 + 2 = 4, or that you are as old as you think you are? How could you tell? In this chapter, we will introduce the study of knowledge and one serious obstacle to the idea that we can know anything, skepticism.

CHAPTER OBJECTIVES

In this chapter, you'll learn about...

- What it means to know something
- Skepticism, an obstacle to knowing
- Arguments against skepticism
- Plato's cave analogy
- The argument from unreliable sense experience
- The dream argument
- The evil genius argument

What Does It Mean to "Know"?

Presumably, you know your birthday. *How* do you know it? There is a wide variety of evidence at your disposal that supports the claim, "I was born on such-and-such date," including the testimony of your relatives, a birth certificate, pictures with the date stamped on them, perhaps (unfortunately) a video, and a birth announcement in the local newspaper. Now, what does it mean to *know* your birth date, that is, what must the world be like for us to say that you *know* when you were born? The branch of philosophy dedicated to answering questions like this is called *epistemology*, or "the study of knowledge." Over the next three chapters, we will introduce you to some of the notable debates in this field.

If it is *you* who knows when you were born, it follows that something about "you" is involved in the process or act of knowing. What part of you? We typically don't think that rocks "know" or that trees "know," so it must be something that distinguishes "you" from these sorts of objects. Similarly, it is not your arm that knows, or your leg—it's *you* who knows. What distinguishes "you" from your arm or your leg? Philosophers have long argued that it is your "mental states" that identify you as you. It is difficult to say just what a "mental state" is, but, minimally, it seems safe to say that mental states are states of conscious thought, such as hoping, worrying, wondering, doubting, believing, loving, etc. They are distinguished, by most philosophers, from "physical states," which include being a certain height, having a certain type of brain, being a certain age, having a certain blood type, etc.

But not just any mental state is sufficient for saying that you know something. It seems you can know a claim without hoping, worrying, or doubting it. However, it doesn't seem that you can know a claim without *believing* it. Believing involves giving your assent to a claim, accepting it *as true* (even if it isn't). So, you may believe a claim without knowing it (e.g., you could *believe* that there are extraterrestrial beings, but not *know* that there are extraterrestrial beings), but you could not know a claim without believing it.

Therefore, we will call believing a claim, a "necessary condition" for knowing— you cannot know without it. A *necessary condition* is a state of affairs that must exist in order for some claim about that state of affairs to be true. For instance, in order for the claim, "This match will light when struck" to be true, there must be oxygen in the room. Oxygen is a necessary condition for fire. But oxygen is not all that is required for a match to light. Oxygen is a necessary condition, but not a "sufficient condition" for the claim, "This match will light when struck." It must also be the case that: The match is dry; it is struck with sufficient force; it is struck against the right sort of surface; etc. Each of these is a necessary condition for lighting a match, and only when they all occur together are they sufficient.

So, while believing is a necessary condition for knowing a claim, it is not a sufficient condition. Believing is not all there is to knowing—as we just saw with extraterrestrials, you could *believe* they exist without *knowing* they do. So, what else is needed? Often when we say we "know" a claim, we mean we are "fully convinced it is true"; for example, "I just know she's cheating on him," or "I know that Hindus are wrong about reincarnation." But when philosophers talk about knowledge, they typically mean something narrower; that is, when someone knows something, they "know it to be *true*." You can be fully convinced of something that is false (e.g., that the Sun travels around the Earth, that atoms look like very tiny marbles, that Santa Claus exists, etc.), but it would seem strange to say you *know* that Santa Claus exists, if, in fact, he doesn't. Therefore, philosophers traditionally consider "truth" another necessary condition for knowing.

So, we have two necessary conditions for knowing: to know a claim, it must (a) be believed, and (b) be true. Is this all there is to it? In other words, are (a) and (b) *sufficient* for knowledge? If I believe that Santa Claus does not exist and he doesn't, do I know that he doesn't? The philosopher Plato (c. 427–347 BCE) didn't think so. Consider the following example.

Imagine you are a juror for a murder trial. The suspect happens to be guilty— the claim, "Joe Schmoe killed Jane Doe," is true. In addition, no one was around when he did it. You listen to all the evidence and find none of it convincing. Nevertheless, you form the belief that Joe Schmoe is guilty. Maybe you do not like the way he looks or how he's dressed. Now this case meets both of the conditions we have identified: (a) the claim is believed by you, and (b) it is true. Do you *know* that Joe Schmoe killed Jane Doe? It doesn't seem so. Your belief isn't connected with the event in the right sort of way. Something is missing.

In his dialogue, *Theatetus*, Plato argues that what is missing is *evidence*; in addition to true belief, you need some *reason* for thinking that Joe Schmoe is guilty before we can say you know it. If his fingerprints were found on the murder weapon, his footprints were found in the dirt near the body, his DNA were found on the corpse, or a witness were to testify to seeing the crime, you would not only have a true belief, you would have a *justified* true belief.

Therefore, philosophers have traditionally argued that to know a claim requires that three conditions are met (that is, they are necessary). Together, these conditions are *sufficient* for your knowing a claim: (a) the claim is believed by you, (b) the claim is true, and (c) the claim is justified for you, that is, you have evidence that it is true. You have a "justified true belief," and this is the traditional definition of knowledge.

An Obstacle to Knowing: Skepticism

Now we have a clear idea of what it means to "know" a claim. So, what sorts of claims, if any, can humans know? One of the most difficult obstacles facing a systematic study of human knowledge is the problem of skepticism. *Skepticism* is the view that our knowledge of reality is limited in some significant sense. Skeptics do not argue that humans do not or cannot have useful beliefs. They don't even argue that we have knowledge at all. They argue, primarily, that humans cannot be justified in holding many of our significant beliefs—we lack evidence sufficient for believing (in the right sort of way) that certain claims (maybe all) are true. Therefore, they challenge the idea that the third condition of the traditional definition of knowledge—justification—can be met.

There are a variety of skeptical views, and each expresses a different sense in which our knowledge is limited. The two most prominent versions are *Pyrrhonian skepticism* and *Academic skepticism*. Pyrrhonian (or "Pyrrhic") skepticism is named after the ancient Greek ruler, Pyrrhus of Epirus (319/318–272 BC), who argued that "things are equally indifferent and unstable and indeterminate" (transmitted by Aristocles, second-century CE). Perhaps the most famous Pyrrhonian skeptic is the Roman philosopher Sextus Empiricus (c. 160–210 CE), who, in his *Outlines of Pyrrhonism*, defends Pyrrhonian skepticism against the followers of Plato (whom he labels "Academics") and the followers of Aristotle (whom he labels "dogmatists").

> **BOX 3-1** Pyrrho and Pyrrhonian Skepticism
>
> It is likely that Pyrrho was not actually a Pyrrhonian skeptic, except perhaps as a practical philosophy. But the tradition handed down to us (through Timon, a student of Pyrrho's, and Aristocles) suggests that Pyrrhonian skepticism was Pyrrho's approach to reality.

Pyrrhonian skepticism is the view that humans do not know whether they have or can have knowledge. It is a position—perhaps it would be more accurate to call it an "attitude"—of extreme doubt; they would not even accept the claim,

"Pyrrhonian skepticism is true." They would just say, "We can't know whether any claims are true; though we call ourselves 'Pyrrhonian.'" Pyrrhonians are not content with their ignorance, however; they are willing to consider arguments showing that Pyrrhonian skepticism is false and that we know some claims about reality. Interestingly (and, for many of us, frustratingly), their arguments against our ability to know are quite powerful, as we will see.

TERMS: Pyrrhic

You may already be acquainted with the term "Pyrrhic" and not even realize it. The Trojan War is often called a "Pyrrhic victory" for the Spartans. The battle was so costly, in lives and resources, and the spoils so few, that it was difficult for them to revel in the victory even though, technically, the Spartans won the war. It didn't *feel* like a win.

The militant ruler Pyrrhus is said to have exemplified the attitude that reality is essentially unsettled and indeterminate, and so there are no conclusive answers to any question. Therefore, a win that doesn't feel like a win is known as a "Pyrrhic victory."

Academic skepticism is not quite as bleak as Pyrrhonism, but it is close. Academic skepticism is the view that human knowledge of reality is highly restricted. The Academics are philosophers who were associated with the ancient philosophical academies, Plato's Academy being the first and most famous. Plato required all students to study ten years of geometry before he would allow them to study philosophy proper. The Academics are distinguished by their philosophical method, often known as the "Socratic Method," still frequently used in law schools. The Academic philosopher asks a student a probing question, and then, in response to each answer, asks question after question, leading the student to what is, apparently, the logical result of his assumptions, which is often absurd. The student is then forced to suspend judgment about his belief until he discovers where he went wrong. Socrates used this method to baffle and confuse some of the brightest minds around him. And though Socrates was Plato's teacher, and therefore not a student at the Academy, he is typically included in the tradition of Academic philosophers.

The Academics are considered skeptics by some because they rarely came to any conclusive conclusions about philosophical topics. Their reasoning led them to suspend judgment, claiming that "[truth] cannot be apprehended" (Sextus

Empiricus, *Outlines of Pyrrhonism*, ch. I, §2). This, however, is a controversial perspective, since there is a rich tradition called "Platonism" that attributes to Plato a set of significant, non-skeptical claims about reality, including, the existence of abstract objects, like numbers and properties (e.g., red, round, rough, justice, beauty), the existence of a creator deity, and a normative political philosophy (see *The Republic*). These beliefs would hardly be characteristic of a skeptic. Nevertheless, you will have to study Plato in detail to determine for yourself whether he was a skeptic.

Some philosophers uncharitably attribute to Socrates a view he did not obviously hold. When some describe Academic skeptics, they say they follow Socrates in claiming, "All I know is that I know nothing." This is uncharitable because Socrates didn't really say this. The line comes at the end of Book I of *The Republic*, where Socrates concludes an extensive discussion about justice by saying: "Then an argument came up about injustice being more profitable than justice, and I couldn't refrain from abandoning the previous one and following up on that. Hence the result of the discussion, as far as I'm concerned, is that I know nothing, for when I don't know what justice is, I'll hardly know whether it is a virtue..." (§354b-c). Socrates is merely saying that, from his discussion with Thrasymachus, he can conclude nothing about the nature of justice.

If Socrates were to make an unfortunate statement like, "All I know is that I know nothing," he would be committing himself to a third type of skepticism, called "universal," or "global," skepticism. *Global skepticism* is the view that humans can know nothing at all. Global skepticism is controversial because it seems impossible for anyone to believe. Consider the following argument. Premise 1 is the definition of global skepticism. We will assume it is true, for the sake of argument:

1. Humans cannot know true claims about reality.
2. Premise 1 is a claim about reality.
3. Therefore, if premise 1 is true and humans can know it, global skepticism is false.
4. And if premise 1 is false, then global skepticism is false (by definition).
5. Therefore, if global skepticism is true, then it is false, and if it is false, then it is false.

The problem is that, if we believe that humans can know nothing, we are accepting as true a claim about the world. If this claim is true and we have good reasons for believing it, then there is at least one claim we know: that humans cannot know any claims about reality. But if this is true, then our starting assumption must be false, because humans can know at least one claim about reality. Any claim that entails its own falsity is called *self-defeating*. Philosophers don't worry too much with self-defeating claims because they cannot be true: assuming they are true for the sake of argument, we can prove they are false.

Therefore, Academic skepticism is only plausible when interpreted as the view that humans can know very little, perhaps only that they can know nothing with certainty, or that they can know only that they know nothing else. Many philosophers that Sextus Empiricus labels as "Academic," including Plato and Carneades, were not obviously committed to global skepticism, though Sextus often writes as though they were.

It is important to note that, unlike Pyrrhonian skeptics, Academic skeptics *do* make claims about reality. They argue that, in fact, we can know little, if anything, from experience, and many argue that we cannot know anything from reason with certainty.

Still Struggling

Explaining a View vs. Defending a View

Up to now, we have only defined skepticism and explained its various incarnations. We have given you no reasons to think any version of skepticism is true or plausible. Philosophy books and articles are often arranged a very specific way to: (a) present someone else's view; (b) argue that that view is implausible; (c) explain the author's own view; and, (d) present an argument that their view is plausible. Be sure you can tell the difference. Below we will give an example, and in what follows after this box, we will present an argument that (a) one interpretation of Academic skepticism is implausible, and (b) that this leaves only one plausible way to understand the Academic skeptic's view.

Example: Presenting the view, "God exists."

"By 'God exists,' I mean that a supernatural being, who is all-powerful, all-knowing, and all-good, and who created the universe out of nothing, exists."

Example: Defending the view, "God exists."

"Something cannot come from nothing. Impersonal objects cannot act without being acted on. And the universe is not infinitely old. Therefore, a being that can act without being acted on, and that was smart enough and powerful enough to bring the universe into existence, must have existed prior to the universe."

Despite their differences, Pyrrhonian and Academic skeptics employ the same strategy against philosophers who claim that humans can know claims about reality. The strategy is simple and quite powerful: present compelling arguments against both sides of an issue. If no evidence is sufficient for believing either view (because both arguments are compelling), the only remaining option is to suspend belief about that issue. Once you've accomplished this for a few foundational beliefs (that our senses are reliable, that memory is reliable, that logic is reliable), then most, if not all, of our beliefs are called into doubt.

In the next section, we will explain four arguments for skepticism. These arguments are aimed at undermining the belief that we know pretty much what we think we know about reality. The first three are aimed at experiential evidence (recall from Chapter 2 the distinction between experiential and non-experiential evidence), and the fourth is aimed at both experiential and non-experiential evidence. If these arguments are successful, we have very strong reasons to doubt that we can know anything about reality.

Four Arguments for Skepticism

1. Plato's Cave

The most dramatic ancient example of an argument for skepticism is found in Plato's famous book, *Republic* (Book VII), in what is known as "The Allegory of the Cave." Plato, speaking through Socrates, asks us to imagine a group of people born and raised in a cave, shackled in such as a way that they can look only straight in front of them at the back wall of the cave. Imagine also that, behind them, there is a path leading up an incline to a mound. Behind the mound, a large fire is burning. On top of the mound, people (puppeteers) are walking back and forth, carrying all sorts of objects: "statues of people and other animals, made out of stone, wood, and every material. And, as you'd expect, some of the carriers are talking, and some are silent" (514b–515). Since the people below are shackled and can look only at the back wall, all they ever experience are the shadows of the objects that pass in front of the fire, and never the actual objects. And they are even further removed from the actual objects the statutes represent! All they know of reality through their senses is what they can observe of the shadows and the sounds made by the carriers. "Then the prisoners would in every way believe that the truth is nothing other than the shadows of those artifacts" (515c).

Imagine, also, what might happen if one of the prisoners were suddenly freed from his shackles and forced into the real world outside the cave. Plato doesn't explain how this might happen, but simply says: "Consider, then, what being released from their bonds and cured of their ignorance would naturally be like…he'd be pained and dazzled and unable to see the things whose shadows he'd seen before" (515c). After his eyes adjusted, the freed prisoner would learn all the aspects of reality he had been missing. And if he tried to return and tell his fellow prisoners, they would not believe him, would ridicule him, and perhaps even kill him. Plato explains, "Wouldn't it be said of him that he'd returned from his upward journey with his eyesight ruined and that it isn't worthwhile even to try to travel upward? And, as for anyone who tried to free them... wouldn't they kill him?" (517a).

In Plato's story, someone is freed from the cave to discover what reality is actually like. In addition, Plato seems to imply that this is not something the prisoner could do on her own, or would welcome, if offered. She would have to be freed by some outside source and forced to walk up the path and out into the real world. The question the reader is left to ponder is: How could we ever know whether we are in a cave? And, if we could know, how could we ever get free? If we cannot know anything true of reality while trapped in the cave, and if we would reject or kill anyone who claims to have knowledge beyond the cave, how could we ever obtain knowledge? If there is no compelling answer to this question, it seems the only rational response is to suspend judgment about all our beliefs about reality derived from our senses.

The skeptical argument goes like this:

1. If I can derive any knowledge from my senses, I am not in cave-like circumstances.

2. <u>I cannot know whether I am in cave-like circumstances.</u>

3. Therefore, I cannot know whether I can derive any knowledge from my senses.

Still Struggling

Explaining a View vs. Accepting a View

Just because a philosopher presents an argument, doesn't mean he or she accepts that argument. Many times philosophers present arguments from others, or make up plausible views contrary to their own, in order to explain their faults.

In this chapter, we will see that Descartes offers a number of arguments defending skepticism. But he is not a skeptic! Descartes's goal is to show that skepticism is false and to defend the idea that we can know a substantial number of claims about reality, in order to "provide a solid foundation for science." Nevertheless, he must show that the traditional skeptical arguments will not work. So, he first has to explain them and their rational force. In the next chapter, we will see that Descartes goes on to argue that we have reasons to reject skeptical arguments.

2. The Argument from Unreliable Sense Experience

A second, more intuitive skeptical worry threatens our faith in our senses. How do we know that the perceptions given to us by our senses reflect the way the world really is? Sextus Empiricus lists a number of different reasons for doubting that our senses accurately reflect the world outside our perceptions. For instance, we notice many differences among humans: "The body of an Indian differs in shape from that of a Scythian...." These differences give us reason to believe that our sense faculties may be constructed differently: "Indians enjoy some things, our people other things, and the enjoyment of different things is an indication that we receive varying impressions from the underlying objects."

There are also disagreements among our own senses: "Thus, to the eye paintings seem to have recesses and projections, but not so to the touch. Honey, too, seems to some pleasant to the tongue but unpleasant to the eyes; so that it is impossible to say whether it is absolutely pleasant or unpleasant." If you cool your left hand and warm your right, then place both in the same water, the water will feel warm to your left hand, but cool to your right. Which measurement of texture, pleasure, or temperature is the accurate measure? Which sense faculty should we privilege above the others?

You may object, at this point, that there are objective measures, such as microscopes and thermometers, according to which we can resolve these discrepancies. But why consider one measure, which is also judged by one of your sense faculties, to be more reliable than another?

First, consider why you should believe anything your senses tell you. Presumably it is because you have experienced a degree of stability over your sense experience—things that you perceive to be red or round or soft continue to be so over time. But this means that you do not rely on a single perception to determine your beliefs. Imagine you are wandering through Australia before studying any of the local animals. Suddenly, out of the corner of your eye, you catch a brief glimpse of a furry creature that seems to have a tail like a beaver, a bill like a duck, and feet like a turtle. But you only see it once, and only for a second. Do you suddenly believe there are creatures like this or do you cautiously consider that your eyes are playing tricks on you? Probably you will suspend judgment until you have more evidence. So, in order to believe something on the basis of your senses, you have to perceive it to be a certain way more often than not.

Now, consider the shape of a table top. Perhaps you have considered that it is rectangular. But why? You have never seen it as rectangular? If you don't believe us, try to draw what you see when you see the table. You will never draw a rectangle. If you are honest, you will draw something that looks like one of the following (Fig. 3-1):

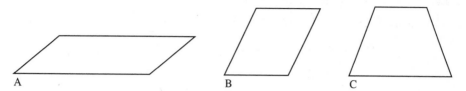

FIGURE 3-1 · Examples of perspective.

To accurately draw objects, you must use the artistic technique called "perspective," that is, you must draw it the way you see it. In this case, you have never seen the table as rectangular. So, why believe it?

Perhaps you believe it because it is useful to believe. But usefulness does not entail truth. Scientists use Newtonian formulas because they are useful, but Einstein showed that they are false. Believing in karma may be useful for coping with reality, but it remains to be shown whether it is exists. Perhaps you believe the table top is rectangular because you can measure it. But presumably you measure it by getting close and standing over it, where it looks more rectangular. But why think this perspective is the most accurate from which to measure the table? Why not stand across the room? If you did, it is likely that your measurements would correspond more closely to figure A, B, or C, than to a rectangle. And why trust your eyes to measure? Surely, your body is a much more useful standard, since knowing that the table is 40" long won't help you avoid running into it, but feeling it would.

French Pyrrhonian Michel de Montaigne (1533–1592) offers similar examples of how our senses deceive us, plus some additional reasons for doubting them. One of the strongest influences on our senses is emotion. Montaigne explains:

> What we see and hear when we are transported with passion, we neither see nor hear as it is…the object we love appears to us more beautiful than it is…and that we hate, more ugly. …To a discontented and afflicted man, the light of day seems dark and overcast. Our senses are not only corrupted, but very often utterly stupefied

by the passions of the soul; how many things do we see, that we do not take notice of, if the mind be occupied with other thoughts? (*Apology for Raymond Sebond*).

As many of us know, during the first few weeks of a new relationship, our significant other can do no wrong. After two months, we suddenly perceive things differently. Even the Greek poet Homer wrote that, "The minds of men are dark or serene, as the day is fair or foul" (*Odyssey*, XVIII).

The skeptical argument goes like this:

1. If I can derive any knowledge from my senses, they reliably represent reality.

2. There is no privileged perspective from which to evaluate the reliability of my senses.

3. Therefore, I do not know whether my senses are reliable.

4. Therefore, I do not know whether I can derive any knowledge from my senses.

3. The Dream Argument

One of the most worrisome of all Pyrrhonian arguments for skepticism is also found in Plato's writings, in a dialogue called, *Theatetus*. Socrates and Theatetus discuss whether it is possible to distinguish dreaming life from waking life. They are not optimistic about the prospects:

SOCRATES: There's a question you must often have heard people ask—the question what evidence we could offer if we were asked whether in the present instance, at this moment, we are asleep and dreaming all our thoughts, or awake and talking to each other in real life.

THEATETUS: Yes, Socrates, it certainly is difficult to find the proof we want here. The two states seem to correspond in all their characteristics. There is nothing to prevent us from thinking when we are asleep that we are having the very same discussion that we have just had. And when we dream that we are telling the story of a dream, there is an extraordinary likeness between the two experiences.

SOCRATES: ... Indeed we may say that, as our periods of sleeping or waking are of equal length, and as in each period the soul contends that the beliefs of the moment are preeminently true, the result is that for half our lives we assert the reality of one set of objects, and for half that of the other set. And we make our assertions with equal convictions in both cases (158b-d).

A significant portion of our lives are spent sleeping. A significant portion of this time is spent dreaming. When we are dreaming, often, we take our dreams to be reality; when we wake up, we take our waking life to be reality. They cannot both be accurate representations of reality. So, which is which?

French philosopher René Descartes (1596–1650) offered a more detailed version of the Dream Argument:

> …here is the fact that I am here, seated by the fire, attired in a dressing gown, having this paper in my hands…And how could I deny that these hands and this body are mind…[?] At the same time I must remember that I am a man, and that consequently I am in the habit of sleeping, and in my dreams representing to myself the same things or sometimes even less probable things, than do those who are insane in their waking moments. How often has it happened to me that in the night I dreamt that I found myself in the particular place, that I was dressed and seated by the fire, whilst in reality I was lying undressed in bed! ("Meditation 1," *Meditations on First Philosophy*).

The problem is this: Even though there are times when I know I am dreaming, there are times when dreaming seems like reality. When I seem to be awake, this could be another delusion of my dreams. Therefore, even though sometimes I can tell that I am dreaming, there is no time that seems like I am awake when I could be *certain* that I am not dreaming.

This is a particularly powerful skeptical argument because for any objection (I can't see people's faces in my dreams; I can't see in color in my dreams; I can't read in my dreams), the skeptic can reply, "Your dreams vary widely, some of this very realistic, some not so much. This could just be a dream where you can do all those things." Descartes concludes, "I see so manifestly that there are no certain indications by which we may clearly distinguish wakefulness from sleep that I am lost in astonishment."

So, the skeptical argument goes like this:

1. If I am dreaming, then all my sensory beliefs are false (because I'm really in bed!).

2. So, if I can derive any knowledge from my senses, then I am not dreaming.

3. <u>I cannot know that I am not dreaming.</u>

4. Therefore, I cannot know whether I can derive any knowledge from my senses.

BOX 3-2 Montaigne's *Apology for Raymond Sebond*

After the fall of Rome and before the Enlightenment, religious and political authorities determined which claims were to be accepted as "knowledge." The most influential of these authorities was the Roman Catholic Church. The Church taught that knowledge of reality is possible because God designed us to perceive it accurately, and that many significant truths of reality must be revealed by God, and the Church is the recipient of this revealed knowledge.

During the Enlightenment, intellectuals began to challenge many of the teachings of the Church. French philosopher Michel de Montaigne, in his letters and essays, revived the arguments of the ancient Pyrrhonian skeptics, challenging not simply the complicated doctrines of the Church, but the ability to know anything at all. Montaigne's most comprehensive defense of Pyrrhonian skepticism is found in his *Apology for Raymond Sebond* (1575–1580). You should also know that the term "apology" is a legal term meaning "defense," and comes from the Greek word, *apologia*.

4. The Evil Genius Argument

Up to now, our skeptical arguments have been aimed at what we can know from our senses. Descartes argues that, even if these arguments are successful, neither Academic nor Pyrrhonian skepticism is justified, because there are a whole host of things we can know that have nothing to do with our senses. He explains:

> …Arithmetic, Geometry and other sciences of that kind which only treat of things that are very simple and very general, without taking great trouble to ascertain whether they are actually existent or not, contain some measure of certainty and an element of the indubitable. For whether I am awake or asleep, two and three together always form five, and the square can never have more than four sides… ("Meditation I," *Meditations on First Philosophy*).

Descartes's point is that we know all the truths of mathematics and logic without appealing to any sense experiences. We do not attempt to justify 2 + 2 = 4 by counting objects over and over, and then drawing an inductive inference about the probability that adding two things to two things yields four things. We know it is true for certain by simply consulting our own minds on the subject. Once we understand what "square" means, we know for certain that it cannot have more than four sides.

But then Descartes does something unexpected. He offers another skeptical argument—an argument no one has ever offered before, and that is even stronger than any that came before it. This argument challenges even our beliefs about logic and mathematics. But, wait; Descartes was no skeptic! He even explains that his purpose is to provide a solid foundation for science. So, what's going on here?

Recall, from earlier, the distinction between explaining a view and accepting a view. Descartes is trying to figure out whether we can know anything at all, so if it is possible to construct another skeptical argument against his own view, even if he has to construct it himself, he has to consider it in order to know for sure.

Descartes begins by saying that he's always believed in an all-powerful God, who is all-good. But he can imagine that he's wrong, and that, instead there is a nefarious deity, bent on deceiving humans. This seems clearly possible—it entails no contradictions—and if it is true, then this evil genius could deceive us about our sensory experiences as well as our mathematical and logical beliefs. If this being is all-powerful but devious, he could make it seem like $2 + 2$ couldn't be anything other than 4, even if, in reality, it is 3 or 5 or 500.

So, the skeptical argument goes like this:

1. If I know mathematical and logical truths, then I am not being deceived by an evil genius.
2. <u>I cannot know I am not being deceived by an evil genius.</u>
3. Therefore, I cannot know whether I know mathematical and logical truths.

This argument casts doubt on all our sensory beliefs, including the belief that we have hands and bodies, and that we have the parents we do, and that we live on a "planet" called "Earth," etc. It also casts doubt on our must fundamental logical and mathematical beliefs, including the belief that $2 + 2 = 4$, that the square root of 25 is 5, that everything is identical with itself, that there could be no married bachelors or round squares, etc. So, if there is no serious philosophical response to this argument, *the Pyrrhonian and Academic skeptics win.*

BOX 3-3 Descartes in Film

Descartes's evil genius argument was popularized in the film *The Matrix*. The main character, Neo, discovers that all his beliefs may be false because he is merely an energy-producing body in the computer system of a group of malevolent machines. Through a series of incredible events, Neo is finally convinced that his whole life had been the product of a computer program, and that now he is in the real world fighting the machines which hold his race in captivity.

Unfortunately, *The Matrix* doesn't highlight the fact that the skeptical problem is not really resolved by Neo's learning that his life has been a computer program—the problem merely becomes obvious. He should now ask whether his new world is the real world or merely the product of some malevolent deceiver. His old world was seamless, so he had no reason to believe he was deceived. Now he knows that world was an illusion. But what about the world outside the matrix? Is it any more real? How could Neo be sure? He couldn't; and now he has more reason than ever to doubt that his beliefs are true. Sadly, seeming to escape one evil-genius world doesn't weaken the problem of skepticism... and it may make it worse!

But we know Descartes is not a skeptic. So, how does he respond? We will explain that in the next chapter. But before you turn the page, look back at the evidence section in Chapter 2. Knowing what types of evidence are available and what types of evidence the evil genius argument causes us to doubt, what options are available to Descartes? Can you think of a response to the evil genius argument?

Still Struggling

The Dream Argument Is Not the Evil Genius Argument

Students of philosophy often conflate the *dream argument* for skepticism with the *evil genius* argument for skepticism. But they are very different. They work the same way: Each argument gives you reason to doubt your beliefs by showing that you cannot distinguish between true beliefs and false beliefs. But the dream argument is only effective against sensory beliefs. Your mathematical and logical beliefs do not change in dreams, therefore, there is no reason to believe that, just because you may be dreaming right now, your belief that $2 + 2 = 4$ is false. But the evil genius argument works even against these. Perhaps the evil genius has convinced you that $2 + 2 = 4$ when, in fact, there are no true mathematical claims at all, or $2 + 2 = 5$. So, keep in mind:

The evil genius argument is not the dream argument!

QUIZ

1. A state of affairs that must exist in order for some claim about that state of affairs to be true is
   A. a necessary condition.
   B. a condition.
   C. a sufficient condition.
   D. an event.

2. What conditions are sufficient for your knowing a claim?
   A. The claim is believed by you.
   B. The claim is true.
   C. The claim is justified for you, that is, you have evidence that it is true.
   D. All of the above.

3. Skeptics do not argue that humans do not or cannot have useful beliefs.
   A. True
   B. False

4. The Academic philosopher asks a student a probing question, and then, in response to each answer, asks question after question, leading the student to what is, apparently, the logical result of his assumptions, which is often _____.
   A. true
   B. debatable
   C. false
   D. absurd

5. Any claim that entails its own falsity is called
   A. self-defeating
   B. self-debasing
   C. self-contrary
   D. none of the above

6. Who lists a number of different reasons for doubting that our senses accurately reflect the world outside our perceptions, such as "The body of an Indian differs in shape from that of a Scythian...."?
   A. Plato
   B. Socrates
   C. Sextus Empiricus
   D. Empiricus

7. Even though there are times when I know I am dreaming, there are times when dreaming seems like reality. When I seem to be awake, this could be another delusion of

    A. my senses.
    B. my dreams.
    C. neither of the above.
    D. both of the above.

8. Descartes offers a skeptical argument—an argument no one has ever offered before, and that is even stronger than any that came before it. This argument challenges even our beliefs about

    A. dreams.
    B. perceptual awareness.
    C. reality itself.
    D. logic and mathematics.

9. In which argument do we find a nefarious deity, bent on deceiving humans?

    A. the dreaming argument
    B. the Pyrrhonian argument from the senses
    C. the evil genius argument
    D. none of the above

10. Descartes was not a skeptic.

    A. True
    B. False

4

Responses to Skepticism

Do the skeptics win? Are we rationally compelled to doubt whether we can know anything of importance? Despite the power of skeptical arguments, many philosophers answer "no." In this chapter we will introduce two very different responses to skepticism: René Descartes's, which inaugurates the tradition known as rationalism, and John Locke's, which inaugurates the tradition known as empiricism.

CHAPTER OBJECTIVES

In this chapter, you'll learn about...

- A few responses to skepticism

- Descartes' cogito argument

- Rationalism vs. empiricism

- The importance of empiricism in Western philosophy

René Descartes to the Rescue

In the last chapter, we looked at four arguments for skepticism: the cave argument, the argument from unreliable sense experience, the dream argument, and the evil genius argument. If any of the first three are successful, that is, if there is no rational response to them, we have reasons to doubt any beliefs derived from sense experience. If the fourth is successful, we have a reason to doubt all of our beliefs derived from sense experience *plus* our mathematical and logical beliefs. Thankfully, many philosophers were not content with these skeptical conclusions, and a few developed responses that led to significant advancements in philosophy.

Recall that the Pyrrhonian skeptic claims that we cannot know whether we know anything, and Academic skeptics claim that we know very little—some that we know nothing from sense experience; some that we only know anything with some degree of probability. Therefore, if anyone can show that we know at least one claim for certain, then Pyrrhonian skepticism is false. If anyone can show that we know more than the Academic skeptic claims, then Academic skepticism is false.

To refute the Pyrrhonian skeptic, Descartes needs only one example of something he can know with certainty. This is because the Pyrrhonian's claim is so strong: "It is unclear whether we can know *anything*." Consider the following two claims:

1. **All swans are white.**   2. **Most swans are white.**

What would it take to prove, conclusively, that (1) is false? You would only need one non-white swan. It cannot possibly be true that all swans are white *and* that there is at least one non-white swan. Therefore, if you know there is one non-white swan, you know that not all swans are white, and (1) is false.

What would it take to prove, conclusively, that (2) is false? One non-white swan would not be enough. It is possible, even plausible, that most swans are white *and* that there is at least one non-white swan. To prove (2) false you would need to know that at least half of all swans are non-white ("most" implies more than 50%; thus, you would need to know that at least 50% of swans are non-white to show that "it is not the case that more than 50% of swans are white"). This would be much harder to do, since you would first need to know how many swans exist plus how many are white and non-white.

The Pyrrhonian skeptic's claim is like (1) above. They doubt whether we can have any knowledge. So, it is enough to prove that we can know one thing for certain to prove, conclusively, that this claim is false. You might ask: But why doesn't the Pyrrhonian just weaken his claim and say, "It is unclear whether we can know most claims"? Skeptics could make this move, but they would no longer be "Pyrrhonian." They would be more like Academic skeptics. And if a skeptic accepted this route, he would be obliged to distinguish those beliefs we can know from those we cannot—a much more difficult task, and one that Descartes embarks on after showing that Pyrrhonism is untenable. If Descartes cannot prove that we can know more than one or two claims, perhaps he must rest content with some weaker form of skepticism. On the other hand, if he can prove that we can know most of what we think we believe, then he need not be a skeptic at all. We will explore this question after we look at Descartes response to the Pyrrhonian.

Descartes begins where he left off with the conclusion of the evil genius argument. Assuming, for the sake of argument, that the evil demon argument is conclusive, is any source of knowledge left? Descartes says, yes. What is left is our *immediate perceptions* of what reality is like—how things "seem" to us, even if any beliefs about these perceptions would be false. It may be false that the wall in front of me is red, but it is true that *it seems red to me*. What's more, I am *certain* that it seems red to me. Even if I were to stop *believing* it is red because someone tells me that there is a red light shining on a white wall, the wall would still *seem* red to me.

There is a popular example of this phenomenon called the Müller-Lyer Illusion (Fig. 4-1). Consider the following two lines. Which seems longer?

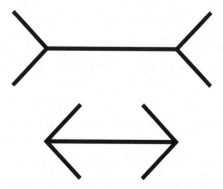

FIGURE 4-1 · The Müller-Lyer Illusion.

To most people, the top line seems longer. However, if you measure the lines, you will see that they are the same length. So, now you no longer *believe* they are different lengths. Nevertheless, they still *seem* different lengths. You have the same perceptual seeming that the lines are different lengths even though you no longer believe they are different lengths. You admit you can be wrong about the actual lengths of the lines. But you cannot be wrong that the lines seem a certain way to you, or so Descartes argues.

Using this seemingly minor philosophical point, Descartes upsets skepticism forever. Remember, the game is to show there is at least one thing that some hypothetical evil genius cannot deceive you about. If this is possible, the Pyrrhonian loses.

The "*Cogito*" Argument

So, in response to the evil genius argument, Descartes argues as follows: The fact that the world *seems to him* a certain way (even if it is false that it is that way) implies that something exists that is experiencing the seeming; that something is Descartes, himself. In order for the evil genius to deceive something, that something must exist. What is it that perceives, however falsely? Descartes. The premises of this argument might look something like this:

1. If I am doubting any of my beliefs, this thing I'm calling "I" must exist to doubt its beliefs.

2. In merely considering the possibility that "I" am being deceived by an evil genius, "I" am doubting my beliefs.

This brings Descartes to his most famous conclusion:

3. Thus, I think (perceive, doubt), therefore I am.*

In Latin, "I think, therefore I am," is, *cogito ergo sum* (kah-**gee**-tō ere-gō **soom**). Because of this, Descartes's argument is widely known as, "the '*Cogito*' argument."

TIP   *Even if Descartes' argument is successful, he proves only his own existence, not yours or mine. All he can know with certainty is that he exists. Maybe you can*

---

*This isn't exactly the way Descartes puts it, though tradition attributes it to him. He actually says something like, "So after considering everything very thoroughly, I must finally conclude that this proposition, *I am, I exist*, is necessarily true whenever it is put forward by me or conceived in my mind." Unfortunately, this is not quite as catchy as, "I think, therefore I am."

know this about yourself, too. But Descartes can't know it about you, and you can't know it about him. If Descartes cannot offer any certain reasons to think anything else exists, he is stuck in a precarious position—he can't know with certainty whether anything else in the world exists except himself. The view that "I am the only thing that exists" is called solipsism. *Some philosophers accept the conclusion of the* Cogito *but argue that Descartes arguments for the existence of an external world are unsuccessful. These philosophers conclude that the only plausible account of human knowledge is solipsism.*

If Descartes is correct, the skeptic loses; there is something each individual person (who exists and can recognize and question her own experiences) can know without doubt: "I exist." Of course, this conclusion doesn't get us very far. We still aren't sure whether we have hands and bodies, or whether other people exist, or whether we are dreaming, or whether the world is how it seems to us. All of these things could still be false if an evil genius exists and is deceiving us. So, in order to accomplish his goal (recall: to provide a firm foundation for science), Descartes must offer reasons for thinking that an evil genius does not exist. And he must start from the only premise he is absolutely certain of: that he exists.

Still Struggling

Imagine you are playing a video game and are in control of a character that you designed. The character exists as a computer program in a computer-generated world. Now imagine that your character becomes conscious of its own perceptions. It now perceives its world as the "real world" and thinks it is a real person rather than merely a computer program—you designed it to think this way and to interact with its computer-generated reality. Now imagine that it questions all of its beliefs about the world—whether it has a body, whether its beliefs about its world are true. It cannot be sure its world is not computer-generated. But if it is conscious of its ability to doubt, it could, using Descartes' reasoning, conclude that it exists (in some form or manner). In fact, it could conclude this with certainty—there is no way it could be wrong. If it is doubting, then, by definition, it exists as a thing. This is the genius of Descartes' argument against the Pyrrhonian skeptic.

From this conclusion, Descartes is able to easily derive two others: "I am a thinking thing," and "I know the contents of my own thoughts." Both of these are implied in the act of inferring that he exists. To derive the conclusion that he exists, he had (a) to be able to consider and doubt any experiences he had from a unique perspective, and (b) to have some experiences that could be deceptive, that is, the world had to *seem* to him a certain way in order for him to ask whether his "seemings" are accurate. If (a) were not the case, then certain thoughts may have existed, like "seeing red now," or "feeling something soft now," but the thoughts would not belong to any one thing—the thought would exist at one moment and not at another. To even entertain whether a thought could accurately reflect reality, something must exist to perceive reality a number of ways. "Seeing a red mug at time 1" could not be false unless there was some awareness of mugs independently of the color red, so that the mug might not actually be red, or might have been another color. This understanding requires that a variety of things, such as mugs and colors and textures, are all perceived from a single perspective, the perspective of a subject, an "I." It turns out that, without this "unity of consciousness," the question of doubt could never arise.

BOX 4-1 CRITIQUE IT!

Twentieth-century philosopher Bertrand Russell argues that Descartes cannot prove that (a) is true. The most that Descartes can conclude from the Cogito is that "*thoughts* exist," not that "*I* exist." Russell explains:

"When I look at my table and see a certain brown colour, what is quite certain at once is not 'I am seeing a brown colour,' but rather, 'a brown colour is being seen.' This of course involves something (or somebody) which (or who) sees the brown colour; but it does not of itself involve that more or less permanent person whom we call 'I.' So far as immediate certainty goes, it might be that the something which sees the brown colour is quite momentary, and not the same as the something which has some different experience the next moment" (1912: 11).

Reread our argument defending (a) and evaluate whether our argument or Russell's is more plausible. What implications would each have for Descartes' argument? For epistemology in general?

Descartes also argues that he cannot be deceived about the contents of his own thoughts. It might be false that the pages of this book are white (since you could be dreaming or in an evil-genius-world, and in both cases, no pages would exist to be white). But it still *seems* to you that the pages of this book are white (unless you have jaundice or a busted blood vessel in your eye, in which case, the pages of this book seem yellow or reddish, respectively). The claim *that* the pages are white (or yellow or red) may be false, but the claim that you see them as white (or yellow or red) cannot be. They either seem this way to you or they don't.

BOX 4-2 CRITIQUE IT!

Twentieth-century philosopher John Pollock argues that Descartes cannot prove that (b) is true. In fact, he argues that, in some cases, our beliefs *about the way reality seems to us* are false. Pollock writes:

"Consider the belief that something looks blue to me. ...[Now] Consider shadows on snow. As every artist knows, shadows on snow are blue, and indeed, upon close inspection they look that way. But many non-artists think that shadows on snow look grey. They hold this as a general belief, presumably inferred from the beliefs that snow is white and shadows on white objects are grey. ... If a person holding such a general belief is asked what color a particular snow shadow looks to him, he may immediately form the belief that it is grey without carefully inspecting his percept. He thus has a false belief about what color it looks to him" ("Procedural Epistemology—At the Intersection of Philosophy and AI," 1999).

Is Pollock right? Would someone who judged shadows on snow believe something false about what he perceives? Or would he just misperceive on the basis of some long-held assumptions? The latter seems unproblematic. If I believe your hair is brown and has always been brown, I may say your hair looks the same, even if you had dyed it a slightly darker shade and even if I would have noticed that had I looked more closely. In this case I misperceived your hair color, though it nevertheless still seemed the same shade of brown it always was.

Is this sort of case a problem for Descartes? If so, what implications would it have for Descartes' argument? For epistemology in general?

Descartes Gets His Senses Back

Let's assume, for the sake of argument, that Descartes is right and there are three claims that he can know with certainty:

1. I exist.
2. I am a thinking thing.
3. I know the contents of my own thoughts.

Descartes now needs to use these beliefs to derive beliefs that will allow him to pursue science. He needs a way to prove that his sense experiences are, in general, more reliable than not.

To do this, he carefully considers what sort of beliefs these three certain beliefs are, that is, what *features* they have that make it impossible to doubt them. He notes that all three seem to be "clear," by which he apparently means "unambiguous." A claim is *ambiguous* if it has more than one specific meaning, for instance: by the claim, "I am going to the bank," someone might mean, "I am going to the financial institution," or "I am going to the river bank," or "I am going to the blood bank." The same goes for claims that contain nouns like, "match," "draw," and "mouse." Descartes could not confuse his own "I" with anyone else's, the word "existence" with any other concept, or the word "thought" with any other idea. Therefore, he concludes that his beliefs are unambiguous.

In addition, Descartes notes these three beliefs all seem "distinct," by which he apparently means "precise" or "not vague." A claim is *vague* if it does not have a precise meaning. Relative terms, such as "tall," "big," "wet," and "hot" often make claims vague. For example, "That is a really small elephant," or "That man is bald." For Descartes, his "I" does not come in degrees, the meaning of "existence" does not change relative to any other state (or non-state) of being, etc. So, he concludes that these beliefs are not vague (and are, therefore, "distinct").

Since his beliefs are both clear and distinct, he reasons, by analogy, that any other beliefs that are clear and distinct must be equally certain. Are there any others? Probably, but Descartes finds one that he can use to his advantage: the idea of "perfection."

Descartes argues that the idea of "perfection" is clear and distinct: it cannot be confused with any other concept, and it has no degrees (anything other than perfection is, by definition, imperfect in some way). Even more interesting, Descartes says that he did not obtain the idea of perfection from the natural world of his senses, since he has never *experienced* any perfect thing, and he did not obtain this idea from his own mind, since *he* is not perfect. So, if he has a clear and non-vague idea that came neither from his experience or himself, it must exist outside himself. Even if an evil genius were to place it in his mind, the evil genius must have gotten the idea from somewhere else, since obviously an evil genius isn't perfect. Where does the idea of perfection come from? Presumably, only from a perfect thing; and the only perfect thing Descartes can conceive of is God.

Now we can add one more idea to the list of things that Descartes can know with certainty:

1. **I exist.**

2. **I am a thinking thing.**

3. **I know the contents of my own thoughts.**

4. *Perfection* **(God)** *exists.*

If a perfect being exists, Descartes reasons, then this being would not allow any evil genius to systematically deceive me about my senses, or allow me to be fooled systematically about whether I am now dreaming. Therefore, if God exists, my beliefs derived from my senses, mathematics, and logic, are, in general, reliable. Does this allow me to believe my senses with certainty? No; I can still be deceived in normal, local ways (objects are not very close, the light is not very good, I am on some medication that makes me hallucinate). And God may allow this to bring about some good in our lives. But, for the most part, my beliefs are true.

BOX 4-3 CRITIQUE IT!

Given these definitions of "clear" and "distinct," can you think of any concepts that are both, but which would still fall prey to skeptical arguments like the dream argument or the evil genius argument?

One example is the concept of "electron." An electron has a very unambiguous, un-vague definition. An electron is an object that has an electrical charge of $-1.602176487(40) \times 10^{-19}$ coulombs. If an object does not have this precise charge, it is not an electron. Therefore, it would seem that "electron" is clear and distinct. Yet, it seems possible that an evil genius has told us this, or that we dreamed it. It might be that actual electrons are very different, or that they do not exist at all.

What implications might this have for Descartes' conclusion that he knows with certainty that God exists?

Are philosophers satisfied with this conclusion? As you may guess, most are not. There are at least two reasons to question Descartes' conclusion. First, Descartes uses the evil genius argument to show that he can doubt even his logical beliefs. But then he seems to "reason" from his belief that he knows the contents of his own thoughts to the conclusion that God exists. Isn't this argument premature? Doesn't it beg the question? (Recall that the fallacy of "begging the question" means to assume something you need to prove.) Doesn't

Descartes need to prove that he can use logic in order to use logic to derive the existence of God? Those who argue that Descartes' reasoning is fallacious, call this the *Cartesian Circle*. In order to reach the conclusion that God exists using logic, Descartes would already need to know that something like God exists to guarantee that his inference is a good one.

Second, even if God exists and Descartes has a noncircular reason for believing it, it may still not follow that God would not allow him to be systematically deceived. Imagine that the world is such a terrible place (because of free will or whatever other justifiable reason) that just waking up in the morning would strike any clear-thinking person with such fear and dread that she would not be able to get out of bed. Perhaps depression would be inevitable and crippling. If this were the case, a good God might choose to systematically deceive humans just to allow them some degree of peace and comfort. As a result, their beliefs about reality would be false, and science would always be inherently severely inaccurate, but people would be, on the whole, happy. We may call this worry, the *Argument from God's Intentions*. In order to reach the conclusion that most of his beliefs are reliable, Descartes would have to know that God desires that our senses accurately represent reality to us. Since it is unclear whether God wants this (even from Descartes' account), his conclusion does not follow.

This leaves us in a nasty place. If the *Cogito* works, each of us can know that he or she exists, that he or she is a thinking thing, and the contents of his or her own thoughts, but nothing else! Some claim that Descartes' discoveries were both a triumph and a defeat. He triumphed over Pyrrhonism, but he left us in the throes of solipsism.

BOX 4-4 Other Important Rationalists

Because Descartes attempted to overcome skepticism using only the evidence available in his mind, he became known as a "rationalist." Rationalists believe that at least some knowledge can be obtained without use of our senses (through reason or logic alone). Philosophers who disagree argue that knowledge is only available through our senses. These philosophers are called "empiricists." Empiricists argue that even logic and math are simply abstractions of language about other beliefs we derived from sense experience.

Other philosophers who attempted to solve skepticism using Descartes' rationalist approach include Nicolas Malebranche, Baruch Spinoza, and Gottfried Wilhelm Leibniz. Malebranche follows Descartes in many respects, but, whereas Descartes claims that ideas exist in a person's mind (modification of the soul), Malebranche argues that they are essentially a part of God. Spinoza uses Euclidean Geometry as a model for investigating knowledge, organizing his arguments into axiom–theorem form, and concludes that our senses are, at best, an aid to genuine knowledge. Leibniz attempts to derive what we can know from two fundamental laws of logic: the law of non-contradiction and the principle of sufficient reason.

John Locke Breaks the Rules

Because of the problems with Descartes' attempt to remove skeptical doubt from science, John Locke (1632–1704) attempted a different sort of argument. Isaac Newton had recently revolutionized the world of physics by offering a theory of motion that was much simpler than previous theories and explained more about how objects move. Locke noted that Newton did not develop his theory using the strategy that Descartes suggested—doubting all he could possibly doubt about physics, and then attempting to build a physical theory on the basis of absolutely certain premises. Beginning with the current theories of motion—those influenced by the work of Galileo, Kepler, and Brahe—Newton compared these according to how objects are observed to act under controlled conditions, and developed a modified theory to account for this motion in light of the standard theories of motion. Newton's new theory explained more of the observable data better than any of his predecessors. Newton's theory was simpler and more powerful than any other theory of physical motion, and therefore, wrought a significant change in the way scientists explain reality.

Taking cues from Newton, Locke decided that if he could observe how people obtain their beliefs, then offer a theory for how this works that is more plausible and powerful than any competing theory, he could bring about a similar revolution in epistemology. This result would show that, while skepticism remains a logical possibility, it is much less plausible than our common sense idea that we know quite a bit about the world.

> **BOX 4-5** Rationalism vs. Empiricism
>
> Notice how Locke's approach now differs from Descartes'. Descartes argues that, because our senses can be deceptive, they should not be trusted as a guide to knowing reality. Instead we should rely solely on the indubitable evidence of our minds. Locke, in stark contrast, argues that we should relax the emphasis on certainty, focus on our senses, and then develop a theory that best explains why we perceive what we perceive. Locke argues that all knowledge is a product of experience in one way or another. This is the view known as "empiricism." So, whereas Descartes' approach to epistemology is rationalist, Locke's is empiricist.

Locke agrees with Descartes that we cannot know with certainty anything beyond what is immediately available in our minds. Nevertheless, he argues that the number of beliefs we can derive from this information is quite extensive. In fact, Locke argues that, even if we cannot know with any degree of precision what the world outside our minds is like, we can know that there is at least something outside of ourselves because our minds are passive recipients of sensation that produces ideas in our minds. Locke defends the idea that our minds are passive recipients by citing examples from our mature adult perception and the perceptions of children. Therefore, something exists outside our minds to produce these sensations that become ideas.

Of course, the question of what produces these perceptions is still open. For all Locke knows, it may be a dream or an evil genius. Is Locke concerned about this possibility? He is not nearly as concerned as Descartes or the skeptics. Locke distinguishes "knowledge" (that which we can know through our senses directly and with certainty) from "probability" (that which goes beyond our immediate senses, but which we may rely on). Locke then argues that, while we cannot know there is an external world, or that our senses are presenting that world to us accurately, we can believe these things are likely to be true. They are probably true. He claims that the primary motivation for skepticism has been a careless disregard for the distinctions between what can be known with certainty and that which is only probable.

But, if he cannot be certain of the beliefs that go beyond immediate perception, how can Locke claim that they are "probable" (i.e., likely to be *true*) in any significant sense? To do this, he first develops a theory about how we obtain our ideas. He says that every belief we have can be traced to two sources of experience: sensation and reflection. Sensation involves the images produced in our minds by the five senses (sights, sounds, smells, tastes, and feels). Reflection involves our perception of the operation of our own minds, that is, how our minds organize and reason about the information received through sensation (perceiving, believing, doubting, reasoning, knowing, willing, etc.). In addition, Locke argues that these are the only sources of information; we receive all ideas after we are born, through experience. We arrive in the world *tabula rasa* (as a "blank slate").

Empiricism Becomes a Plausible Philosophical View

Because Locke believes all knowledge comes through experience; he is an "empiricist." Empiricism stands in stark contrast with rationalism, the view held by Descartes, Spinoza, and Leibniz, who argue that at least some knowledge is obtained non-experientially. To justify his empiricism, Locke has to show that his view is more plausible than rationalism, so he attacks the principle argument for rationalism: Descartes' argument for innate ideas.

After Descartes concludes that he exists, he is a thinking thing, and that he knows the contents of his own thoughts (and before he argues that God exists), he considers what sort of information his thoughts contain. He begins with a piece of wax. At one moment the wax has a certain set of properties: it tastes a little like honey, it smells a little like flowers, it is yellowish, irregular in shape, hard, cold, and makes a sound if you tap it. But at the next moment, after he heats the wax in the fire, it loses all of these properties: It is now tasteless, colorless, shapeless, larger, liquid, hot, and if you tap it, it no longer makes a sound. And yet, he must believe that the second set of properties is a set of the same piece of wax that he observed at the previous moment. "So what was it in the wax that I understood with such distinctness? Evidently none of the features which I arrived at by means of the senses; for whatever came under taste, smell, sight, touch or hearing has now altered—yet the wax remains" (*Meditation 2*).

Descartes concludes that his knowledge of objects (when something is a piece of wax and when it stops being that piece of wax) is a function of his mind, and not a product of his sense experience. Therefore, our ideas of objects are *innate*, given to us in our minds by God prior to our experience of the world. Our senses give us impressions that can be misleading or outright deceitful, but our minds associate sets of impressions (even misleading ones) with objects.

As a sophisticated thinker, you might suppose that Descartes' scientific beliefs were simply too immature to explain why he perceives the wax to be the same thing despite the fact that all of its properties have changed. You might say, "Oh, but not all of its properties have changed. It retains the same physical makeup as before—its fundamental particles have not changed." There are two problems with this response. First, even if you are correct, Descartes cannot perceive these particles, and it is likely that he wasn't fully aware of them, and the point of the thought experiment is to prove that he does not obtain his knowledge of objects from sense experience alone. So, even if you're right, Descartes' conclusion still follows; your ideas about the fundamental particles do not come from direct sense experience.

And second, not all objects are defined in terms of their fundamental properties. Consider Michelangelo's statue of David. Imagine we stole the

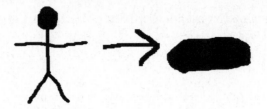

FIGURE 4-2 · The statue of David being crushed to a pile of dust.

statue and crushed it into a billion tiny fragments of stone, carefully pre-serving all dust particles. Now we have a pile of stone, not David, even though we have all the same fundamental particles of the statue of David (see Fig. 4-2).

What's even more worrisome is that, even if, by some artistic miracle, we were able to reconstruct the statue of David from the rubble into a perfect likeness, in size, appearance, and texture, it would be difficult to say that it is, once again, *Michelangelo's* David—it may *look like* it, but now it is *our* statue of David, a cheap (albeit awesome) knockoff of the original.

So, as we noted, Descartes concludes that ideas about objects are not derived from sense experience, but are produced by the mind—they are innate. How does Locke respond to this clever bit of reasoning? Locke argues as follows:

1. If all ideas are innate, then everyone will be aware of a few without expe-rience (e.g., those non-experiential beliefs Descartes was on about: 2 + 2 = 4, everything is identical with itself, nothing can both exist and not exist at the same time, etc.).

2. But children and mentally handicapped people are not aware of these (e.g., try teaching a five-year-old what "identical with itself" means).

3. Therefore, it is more plausible that everyone obtains all their ideas from sense experience.

Since some of the population seem to have no innate ideas (especially children, since we were all children once), it seems more likely that everything we know, all our ideas and beliefs, are derived from the information we receive from our senses. If this is right, Locke's empiricism is more plausible than rationalism.

BOX 4-6 CRITIQUE IT!

Some philosophers and psychologists now suggest that infants, even as young as 5 months, can express an understanding of mathematical, physical, and even moral concepts. In the 1980s, psychologists Elizabeth Spelke and Renée Baillargeon discovered that babies behave much like older humans in response to images that apparently violate the laws of nature, for instance, magic tricks. In 1992, psychologist Karen Wynn discovered that babies react to simple mathematical mistakes much the way older humans do. And in 2009 and 2010, psychologists Paul Bloom, Karen Wynn, and Valerie Kuhlmeier discovered that babies may even possess some rudimentary understanding of moral behavior.

What do these results suggest? Since it is unlikely that these babies have been taught math or physics at 5 months of age, the results suggest that these concepts may, in fact, be innate. If further research bears them out, Locke's premise 2 is likely false, and his rejection of rationalism falls flat.

But there remains the question of how we could know anything about *objects*. Even if Locke is right, we still need to explain why Descartes perceives the wax as the same wax even though all its properties changed, but doesn't perceive David as the same David even though only one property changed (shape).

In response, Locke suggests that Descartes is right that our minds play a role in identifying objects, but Locke argues that minds *produce* the idea of objects from sense experience, rather than receiving them wholesale from the world. Things that we call objects have properties, and Locke distinguishes two types of properties, which he calls "qualities": primary qualities and secondary qualities. *Primary qualities* are properties that an object has independently of our perceiving them: for example, density, mass, volume, speed. Our senses are organized to represent these primary qualities in a certain way. Our eyes are not organized the way an eagle's are, and our ears are not organized the way a bat's are, so it is likely that we see and hear differently than an eagle and a bat. Therefore, the qualities of an object that have to do directly with our senses are actually produced by our senses out of the primary qualities. Locke calls sensory properties *secondary qualities*.

With this distinction between primary and secondary qualities, Locke explains how we obtain our ideas about the world. It is a three step process:

1. An object's primary qualities act on our senses (see Fig. 4-3).
2. Our minds combine these primary qualities with secondary qualities, forming an idea of the object (see Fig. 4-3).
3. The idea of the object is then implanted in our minds (see Fig. 4-4).

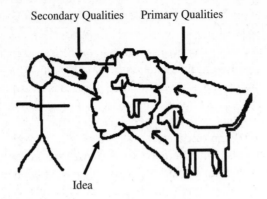

Secondary Qualities   Primary Qualities

Idea

FIGURE 4-3 · Primary and secondary qualities combine to produce an idea of an object.

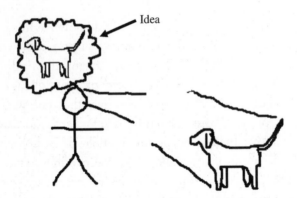

Idea

FIGURE 4-4 · The idea is printed in the mind.

This theory of how we obtain objects became known as *The Representation Theory of Perception*, and it remained popular until the 1950s (not bad for a guy writing in the 1700s). This theory explains how we can be mistaken about objects (some animals' sensory faculties are better suited to the world than ours), and why we would call the different properties of the wax the same piece of wax and yet think differently about statues (our minds organize their primary qualities differently).

It is fine to have a theory about how our perceptions work, but what about the skeptics? Does Locke's theory help respond to skeptical worries? To some degree, yes. We can know much more than the skeptics originally claimed merely by focusing on how we perceive the world. In addition, it is likely that objects exist, having a number of primary qualities, even if we cannot know precisely what these objects are like. However, after several hundred pages, Locke admits: "I am apt to doubt that, how far soever humane [human] Industry [science] may advance useful and experimental Philosophy in physical Things, scientifical will still be out of our reach" (*Essay Concerning Human Understanding*, Book IV, Ch. III, §26). Though Locke relies heavily on probability for making claims about the world outside our senses, he admits that this is not sufficient for establishing an accurate and enduring scientific picture of the world.

Therefore, while Locke advances our understanding of our own minds, he does little to quell broader, perhaps more significant, skeptical worries. We may still be committed to some form of solipsism. And, unfortunately, things get worse before they get better as David Hume calls into question even Locke's reliance on probability. We'll take a look at Hume's response to Locke in the next chapter, along with some contemporary developments in the field.

QUIZ

1. If the dream argument is successful, we have a reason to doubt all of our beliefs derived from sense experience *plus* our mathematical and logical beliefs.
   A. True
   B. False

2. It may be false that the wall in front of me is red, but it is true that
   A. it is still red.
   B. it seems red to me.
   C. both of the above.
   D. neither of the above.

3. One conclusion Descartes draws is
   A. "I know the contents of my own thoughts."
   B. "I know that I do not know."
   C. Neither of these above.
   D. The combination of these above.

4. According to Descartes, to even entertain whether a thought could accurately reflect reality, something must _____ to perceive reality a number of ways.
   A. want
   B. need
   C. begin
   D. exist

5. "I exist," "I am not a thinking thing," and "I know the contents of my thoughts" are claims that Descartes is sure of.
   A. True
   B. False

6. A claim is _____ if it has more than one specific meaning.
   A. equivocal
   B. vague
   C. ambiguous
   D. false

7. According to Descartes, what (a) cannot be confused with any other concept, and (b) has no degrees?
   A. perfection
   B. imperfection
   C. belief
   D. knowledge

8. **According to Descartes, what guarantees that one's beliefs derived from one's senses, mathematics, and logic, are, in general, reliable?**

   A. one's own mind
   B. clear and distinct ideas
   C. clear ideas
   D. God

9. **In order to reach the conclusion that God exists using logic, Descartes would already need to know that something like God exists to guarantee that his inference is a good one. This is known as**

   A. Descartes' Law
   B. The Argument from Circularity
   C. The Cartesian Circle
   D. None of the above

10. **Primary qualities are properties that an object has independently of our perceiving them: for example, density, mass, volume, speed.**

    A. True
    B. False

The Problem of Induction and the Development of Externalism

Although there are powerful (and according to many philosophers, successful) responses to many worrisome types of skepticism, some skeptical doubts linger. One of the most famous skeptics, David Hume, makes the problem of skepticism especially acute with what is now known as the "problem of induction." In this chapter, we will introduce Hume's influential argument, explain some notable responses, and show where these discussions led twentieth-century epistemologists.

CHAPTER OBJECTIVES

In this chapter, you'll learn about...

- David Hume's revival of skepticism
- The problem of induction
- Attempts to solve the problem of induction
- Immanuel Kant's response to Hume
- Logical Positivism
- Gettier-style problems
- The move to externalism

David Hume Revives Skepticism about Science

David Hume, unlike Descartes and Locke, does not begin his study of epistemology with the goal of providing a foundation for science. On the contrary, he aims to show that a robust scientific view of reality is not possible given human limitations. He writes: "Man is a reasonable being; and as such, receives from science his proper food and nourishment: But so narrow are the bounds of human understanding, that little satisfaction can be hoped for..." (*Enquiry Concerning Human Understanding*, 3). Rather than increasing our ability to justify our claims about reality, Hume tears away at some of our most cherished beliefs.

Notably, Hume accepts many of Locke's conclusions. Hume is also an empiricist, so he argues that all we can know is obtained through experience. And like Locke, he distinguishes two sources of experience: matters of fact (which includes all the information provided by our senses) and relations of ideas (which includes how our minds organize and understand matters of fact). As you can see, these categories roughly correspond to Locke's Sensation and Reflection. Hume also distinguishes perceptions (which he calls "impressions") from thoughts about those impressions (which he calls both "thoughts" and "ideas"). With deference to the skeptics, Hume does not presume that impressions are caused by reality external to our minds, and, in fact, he argues that any claim that expresses something beyond what we can immediately perceive is not justified. We do not know whether there is a reality beyond what we can perceive—even outside our immediate perceptions—therefore such a question is meaningless.

In addition, Hume argues that even well-founded conclusions about what we can know through our sense experience are insufficient for justifying even probabilistic inferences about nature in general. He argues that we cannot reasonably make even *probabilistic* scientific claims about the external world.

Most of our claims about reality have little to do with what we directly experience; more often, they are inferences from cause to effect: if I press the brake, the car will slow down; if I travel by plane, my plane won't crash; if I see a cat stick its head around a corner, I infer that the rest of the cat is around the corner; if I am hungry, eating should satisfy me; if I smile at someone, that person will probably smile back; if I'm late for the meeting, I won't get the job, etc. But what supports our inferences from causes to effects?

Consider the example of dropping a pen. If I hold a pen a few feet above the floor and let go of it, will it fall? Most of us believe it would. Do we believe this because we have *seen* this pen fall *in two minutes*? No, two minutes from now is in the future, and none of us have seen anything in the future. Do we believe the pen will fall because similar objects have always fallen in similar circumstances? Maybe; but we only tend to believe that because we also tend to think there are laws of nature that govern the motion of objects. For instance, we would not draw the same conclusion about letting go of a helium balloon a few feet above the floor. Why do we believe in natural laws? We do not immediately experience such a law (we can't touch it, taste it, smell it, etc.), only events that *might* be guided by regularities we happen to call "laws." All we observe are events (letting pens go; pens falling). Why think something like a law exists to connect these events?

Here, Hume does something revolutionary. He introduces an argument that philosophers today are still wrestling to avoid: *The Problem of Induction.* Consider an argument that the pen will fall when I let go of it from the premise that pens always fell when I let them go:

1. All the pens I let go (in the past) fell to the floor.
2. <u>I will let go of this pen in two minutes.</u>
3. Therefore, this pen will fall to the floor in two minutes.

What connects premises 1 and 2 with the conclusion? First of all, we cannot say "induction," since induction is the very type of reasoning we are attempting to prove. Induction is reliable if and only if the future will look like the past. But if we add this as a premise, we make the argument fallacious:

1. All the pens I let go (in the past) fell to the floor.
2. I will let go of this pen in two minutes.
→**4.** <u>The future will look like the past.</u>
3. Therefore, this pen will fall to the floor in two minutes.

This is the very piece of information we are trying to prove. The conclusion is an instance of the future looking like the past, therefore, if we assume that the future will look like the past, this argument is circular (it "begs the question," which means it assumes in the premises what we are trying to prove in the conclusion). So:

(a) **Induction cannot connect the future with the past (Induction merely says that the future is connected with the past; therefore, it is question-begging.).**

In addition, I do not perceive any connection between them:

(b) **Nothing in my sense perceptions connects the future with the past (I do not *see, touch, taste, smell,* or *feel* any natural laws, and I do not perceive what happens two minutes in the future.).**

And third, adding a premise to make the argument deductive also makes the argument fallacious:

1. All the pens I let go (in the past) fell to the floor.

2. I will let go of this pen in two minutes.

→5. If I let go of this pen in two minutes, it will fall to the floor in two minutes.

3. Therefore, this pen will fall to the floor in two minutes.

But this is the very connection that I am trying to prove, therefore, the argument is circular (it "begs the question"). So:

(c) **Deduction cannot connect the future with the past (Deduction would require that we merely restate the connection we need as a premise; therefore, it is question-begging.).**

Hume shows that none of the three ways of establishing the truth of any claim (direct sense perception, induction, and deduction) are sufficient for justifying our belief in induction. Because science depends essentially on inductive reasoning, Hume concludes that we should be skeptical of all scientific claims.

Still Struggling

Hume's main argument can be formulated very simply:

1. We should believe scientific claims only if we have reasons to trust induction.
2. We have no reasons to trust induction.
3. Therefore, we should not believe scientific claims.

The important bit is his argument for premise 2. Why think induction is not reliable? Induction is basically the belief that reality is stable and uniform (it doesn't change much, and it is governed by laws). How do we justify this belief? We do not directly perceive the laws of nature, and any inference about them requires induction—the very reasoning process that is in question. Since there is no way to justify our belief in induction, induction is not a trustworthy method of reasoning.

Now, you might say, "But induction has always worked." True, but this, also, is a claim about the past, and you would need induction to extend it to future events. Imagine that someone is flipping a coin and asks you to guess the result before each flip. Imagine that, for the first 25 flips you guess correctly. Is this a reason to believe that "guessing" is a reliable source of reasoning about coin flips? Surely, not. Hume argues that we all face the same problem with induction; the fact that it has been reliable, is not evidence that it will continue to be reliable. Maybe we've just been lucky.

If Hume is right and the Problem of Induction undermines our ability to reason scientifically, how does he explain our tendency to rely on induction in our everyday lives? That is, why I am I not afraid to cross empty streets, but more cautious when there is traffic? Hume says this tendency is a mere habit of the mind. The fact that induction has been of great use to you has led you to believe that it is a trustworthy principle of reality. And Hume even admits that he cannot live without it in his daily life. He writes:

> ...since Reason is incapable of dispelling these clouds, Nature herself suffices to that purpose, and cures me of this philosophical melancholy and delirium, either by relaxing this bent of mind, or by some avocation, and lively impression of my senses, which obliterate all these chimeras. I dine, I play a game of backgammon, I converse, and am merry with my friends. And when, after three or four hours' amusement, I would return to these speculations, they appear so cold, and strained, and ridiculous, that I cannot find in my heart to enter into them any farther (*Treatise of Human Nature*, Book I, Part IV, Section VII, ¶9).

So, pragmatically, we cannot live without it; philosophically, it is irrational. What are we to do?

Attempts to Solve the Problem of Induction

Thankfully, many philosophers are not content to concede the loss of scientific reasoning. There are a number of serious attempts to solve the problem of induction, and we will briefly discuss three: one from Immanuel Kant, one from Karl Popper, and one from John Hospers.

Immanuel Kant (1724–1804), whom we will discuss further in the next section, offered an interesting response to the problem of induction. Until Kant, philosophers labored under the impression that claims about the natural world, "synthetic" statements, were somehow mutually exclusive with claim that can be known without any experience whatsoever, that is, *a priori* claims. *A priori* is Latin for, "from the previous," and is used in philosophical contexts to mean, "without experience." Prior to Kant, all *a priori* claims were considered to be *analytic*, meaning we can know them solely in virtue of the meaning of the terms involved. For instance, we can know the logical claim that "(A & B) entails A," just by understanding what "&" means. The traditional view is that we can know claims about mathematics and logic *a priori*, but nothing about the natural world; knowledge of the natural world requires experiential, or *a posteriori*, evidence (from the Latin, meaning "from what comes after," and is used in philosophical contexts to mean, "by or through experience"). Kant challenged this distinction, arguing that, while most synthetic claims are known *a posteriori*, some synthetic claims are knowable *a priori*. Among these synthetic *a priori* truths, Kant included, surprisingly to many, arithmetic and geometry.

While pure logic and some abstract mathematical concepts are analytic, there is nothing in the numbers "5" and "7" that would suggest the number "12," so Kant argues, therefore, arithmetic claims require some experience of the world to understand the meaning of "5," "7," and "12." But once this meaning is established, we can understand the relationship between them *a priori*, namely, that "5 + 7 = 12." Therefore, this claim is a synthetic *a priori* claim. He uses the same sort of argument to show that our understanding of some geometric concepts depends on experiencing objects in space, though our understanding of geometric relationships is only attainable without experience, that is, *a priori*.

In response to Hume's problem of induction, Kant argues that our perception of the interaction of objects leaves us with an undeniable impression of a "principle of universal causality," which underwrites our perception of the regularity in nature. As Hume notes, we experience only a series of events, one after the other. Kant goes even further, saying that imagination, not perception, places them in order in time. And because our imagination acts on this series to organize it intelligibly, we have an *a priori* understanding of their causal relationship.

Kant explains:

> Now in order for this [order of events] to be cognized as determined, the relation between the two states must be thought in such a way that it is thereby necessarily determined which of them must be placed before and which after rather than vice versa. The concept, however, that carries a necessity of synthetic unity with it can only be a pure concept of understanding, which does not lie in the perception, and that is here the concept of the **relation of cause and effect**, the former of which determines the latter in time... (*Critique of Pure Reason*, Pt. II, Div. I, Bk. II, Ch. II, B, B234, Trans. Paul Guyer and Allen Wood).

Kant is saying that, in the act of perceiving events, our minds organize them in time as causally related. This causal organization carries with it the impression of necessity. Hume calls this merely a "habit," reflecting nothing more than a prejudice of experience. Kant says it is a synthetic *a priori* law, necessary for us to have any experiences at all. This universal law is prompted by experience, organized by our minds, and is thus, synthetic *a priori*. Therefore, we are fully justified in drawing inductive inferences because we have synthetic *a priori* knowledge of a universal law of causality that underwrites the uniformity of nature.

There are two problems with this approach. First, synthetic *a priori* truths are controversial. Empiricists, classical and contemporary alike, generally reject them, arguing that all of our knowledge is obtained either analytically or experientially. The primary worry is that Kant also thought Euclidean geometry gave us privileged access to the structure of space independently of experience. Unfortunately, Einstein's General Theory of Relativity offers an account of space that depends essentially on a non-Euclidean geometry (Riemannian, to be exact). This runs counter to Kant's claim that we can know a synthetic truth about reality *a priori*, and therefore casts doubt on the whole enterprise. There are some contemporary proponents of something like synthetic *a priori* justification, including Laurence BonJour and George Bealer (2002). Both argue that induction is justified non-experientially, though, to avoid the pitfalls of Kant's account, their views differ widely from Kant's.

Second, even if the universal law of causality could be justified as a synthetic *a priori* truth, it is not clear that it would actually solve the problem of induction. Twentieth-century philosopher of science, Wesley Salmon, explains: "For each occurrence it claims only the existence of *some* prior cause and *some* causal regularity. It gives no hint as to how we are to…identify the causal regularity. It therefore provides no basis upon which to determine whether the inductive inferences we make are correct or incorrect" (*The Foundations of Scientific Inference*, 1966, p. 43). The principle states only that nature is governed by regularities; it doesn't tell us which of the events we perceive are related causally and which are related only accidentally.

A second notable attempt at solving, or rather avoiding, the problem of induction comes from twentieth-century philosopher of science Karl Popper (1902–1994). Popper agrees with Hume that the problem of induction has no rational solution. To avoid the problem, Popper contends, contrary to Hume, that science doesn't need induction after all. Popper argues that the proper method of science is to test hypotheses for disproof, rather than for proof.

No amount of experience of a phenomenon (say, seeing a white swan) can ever guarantee a universal claim about the phenomenon (that all swans are white). Therefore, Popper argues that induction is merely an illusion, a useful fiction, but inappropriate for science. But one experience contrary to a hypothesis (say, one black swan), can disprove a universal claim (that all swans are white). Therefore, scientists form hypotheses and then test them for disproof, a view known as *Deductivism*. A claim is disprovable if it is falsifiable, and all legitimately scientific claims must be falsifiable, that is, there must be some set of circumstances under which they are false. This is not to say that they must be false—only that there must be some way of showing that they are false, just in case they are. An example of an unfalsifiable claim is, "No car runs forever." To disprove this statement, you would have to have a way of knowing that the car exists at every future moment in time—something, presumably, no one could do. On the other hand, "All cars run forever," is falsifiable. One car that no longer runs would disprove it.

But falsifiability is not enough to get science up and running, since, while many scientific claims have yet to be falsified, many claims that have never been tested have not been falsified, either. So we need a way to distinguish "scientific" claims from mere claims that haven't been tested at all. Popper introduces the idea of "corroboration" as a selection criterion. According to Wesley Salmon, Popper "suggests selecting the most falsifiable hypothesis. Thus, he recommends selecting a hypothesis with low probability. According to Popper, a highly falsifiable hypothesis which is severely tested becomes highly corroborated" (*The Foundations of Scientific Inference*, 1966, p. 25). A hypothesis is corroborated if, after repeated testing, it is not falsified. This doesn't mean it is true, but merely that it has survived strict scrutiny. Therefore, Deductivism is a combination of falsifiability and corroboration. If Popper is right, this precludes any need for induction in science.

Unfortunately, however, Popper's view is subject to powerful criticism. Imagine trying to test the hypothesis (H1):

H1: The Earth is flat.

To test this, you form the following test implication (TI):

TI: If the Earth is flat, then if I stand on the shore and watch a sailboat sail out to sea, the hull should not disappear before the mast (see Fig. 5-1).

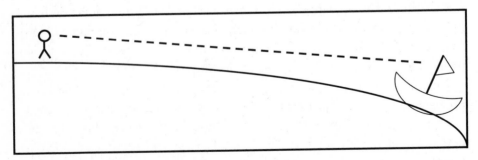

FIGURE 5-1 • If the Earth is flat, the hull will not disappear before the mast.

If, in fact, you watch a sailboat sail out to sea and the hull does disappear before the mast, it would seem that you have falsified the claim, "The Earth is flat." But on closer inspection, you notice that TI includes some fairly significant assumptions:

A1: Light travels in straight lines.

A2: The boat did not sink.

A3: My eyes are working properly.

Each of these claims is a falsifiable scientific claim, and must also be tested as a hypothesis. To draw the inference that the Earth is not flat, none of A1–A3 can be falsified. But, according to Popper, none can be verified, either. Each may be better corroborated than some others (though, thanks to Einstein, we now know that light doesn't always travel in straight lines), but none can be accepted as true, or even probable, since to accept a universal claim as true or probable would require induction. Therefore, in order to disprove H1, Popper must be able to prove A1–A3. Since his system prevents him from proving A1–A3, Popper cannot decisively falsify H1. Since every testing scenario requires some scientific assumptions, Popper's strategy for avoiding the problem of induction seems doomed.

TIP   *There is often some confusion on this point about probability. Probability is always forward-looking, never backward-looking. That some medicine has cured my headache 75% of the time in the past is a "statistic," a claim about the past. To reason from this statistic that the medicine is 75% likely to cure my headache next time is a probability, a claim about the future. By definition, then, probability claims require induction.*

Finally, a third strategy for solving the problem of induction comes from contemporary philosopher John Hospers (b. 1918). Hospers agrees with Hume and Popper that we cannot prove induction on rational grounds. Nevertheless, he argues that there are prudential reasons for accepting it *as* true. First, without induction, the word "evidence" would lose any semblance of meaning. Hospers explains:

> A million times in the past when I have let go of a book or a stone it has fallen; never once has it risen into the air. This, and all the other things I know about the behavior of material objects, lead me to believe that it will fall when I let go of it this time. Is there really *no* evidence either way as to what the stone will do next time? To say this is to abandon all prediction and all science, and, more important here, *all meaning for the word "evidence"* ("A Pragmatic Solution to the Problem of Induction," in Rauhut, *Readings on the Ultimate Questions*, 2nd ed. New York: Pearson-Longman, 2007, p. 103, italics his).

Hume might be perfectly content that, philosophically, the problem of induction leads us to abandon all prediction and science, but it is unclear whether he would be so content to lose the word "evidence." Why does Hospers think giving up induction is tantamount to giving up evidence? If we take evidence to mean, "a reason to believe *X*," then even Hume agrees that, in his everyday life, he takes past experience as evidence for the truth of future claims. So, if we're not willing to give up the concept of evidence, and evidence is inextricably tied to induction, then we have a pragmatic reason—it is *prudent*, or in our best interests—to believe in induction.

Second, not all beliefs require rational justification. Axiomatic systems, such as arithmetic, geometry, and logic, must assume their first principles. The first principles of any system cannot be proven within that system, on pain of circularity. But we do not give up math or logic merely because we cannot prove their axioms independently of their theorems. Induction is as fundamental to our scientific reasoning as Euclid's Five Postulates are to Euclidean geometry. Therefore, if we are not willing to give up math or logic because they depend on unprovable first principles, neither should we give up induction as a first principle of science.

BOX 5-1 CRITIQUE IT!

Hospers' pragmatic solution is interesting because it forces us to seriously consider our understanding of evidence and the beliefs we derive from axiomatic systems (e.g., "$2 + 2 = 4$," and "If A and B, then A"). For practical purposes, we could not give up most of these beliefs, but if we are being honest about our rationality, as Hume asks us to, have we really "justified" our belief in evidence, math, or induction? Hume, of course, would say no.

To be sure, we experience plenty of regular coincidences (every time I walk into my office building, the lights are on), and it is difficult to distinguish these from causal regularities (every time I pull on my office building door, it opens). But there is one question that Hume can't seem to answer. Why are there any regularities at all? What explains the apparently law-like relationships among events that we live by—and that we exploit to remain alive?

Hume surely cannot say "chance." Some events in the past have been too precisely related too many times to be a product of mere chance (we wrote laws describing them). But, similarly, surely Hospers cannot rest content that they are *merely* "useful." Something must explain their usefulness. So, to claim that induction is a matter of chance is as irrational as believing the conclusion of a circular argument. And to believe that induction is useful without believing that anything *explains why* it is useful is also irrational. Therefore, it might seem that, not only is it not irrational to believe in induction, the claim that nature is uniform may be the *best explanation* for many of the regularities we have observed in the past.

Hume's Aftermath: From Kant to the Twentieth Century

Responses to Hume dominate the philosophical landscape from the 1700s through the twentieth century. Hume laid the groundwork on which other Scottish philosophers built, including contemporaries Thomas Reid (1710–1796), who objected to Hume's account of causation, and George Campbell (1719–1796), who objected to Hume's argument against miracles. Almost every major philosopher since the 1700s has had to face one or more of Hume's criticisms. But the most influential philosopher motivated by Hume's skepticism, the philosopher who brought about the next major revolution in philosophy, was Immanuel Kant.

BOX 5-2 Another Important Empiricist

Irish Anglican Bishop George Berkeley agrees with Locke and Hume that we can only obtain knowledge through experience. Berkeley, however, offered a much more radical explanation for how we obtain knowledge than Locke or Hume. Like Descartes and Locke, Berkeley argued that all we have direct access to are the images in our minds. The only way images can be produced in our minds is for something to cause them. But, in contrast to Descartes and Locke, Berkeley argues that contingent objects or events (things that depend on other things for their existence, namely ideas and material substances) cannot cause other contingent objects or events. In fact, the only possible causal force is a necessary being (a spiritual substance), namely, God. Everything that exists, including us and the images in our minds, is a product of God's imagination, and every event is directly produced by him. Therefore, although Berkeley is an empiricist, the only things that exist are persons and minds—no external world, no objects beyond mental objects. So, there really are tables and cats, but tables and cats are really just ideas in the mind of God.

This account is a philosophical position known as *Idealism*. Other idealists include Fichte, Hegel, and J. M. E. McTaggart. Many idealists tried to remove the notion of God from their views. Unfortunately, without a divine being that can act freely, there is no explanation for the regularities in nature. On the other hand, if God is the ultimate source of all events, it is not clear how humans might be morally responsible for their actions or why any immoral actions exist.

Immanuel Kant: Awakening a Giant

Immanuel Kant lived in Prussia at the end of the eighteenth century and stud-ied philosophy at the University of Köningsberg, where he came to respect very much the work of the German rationalist Christian Wolff. But when Kant read Hume, he tells us that Hume "interrupted my dogmatic slumber" (*Prolegomena to Any Future Metaphysics*, Preface). Hume presented such a challenge to the rationalist picture of epistemology that Kant spent the rest of his career attempt-ing to resolve the problems Hume had either created or exacerbated.

We do not have the space to walk through each of Kant's arguments, but the result of his work caused as much of a stir in philosophy as Descartes' *Cogito*. Rather than beginning with epistemological questions about what we can know, Kant begins by asking what reality would have to be like for us to even perceive anything—what are the necessary conditions for having any thoughts at all. He introduces an approach to philosophy that he calls "transcendental," which he intends to be distinct from both epistemology and metaphysics. Today, we might call his project "meta-metaphysics" or "meta-epistemology," or some combination of the two. Kant's goal was to produce some sort of middle ground between rationalism and empiricism, and to thereby forego the skeptical conclusions of Hume.

The elegance of Kant's approach is that it includes detailed arguments about the way reality would have to be (on pain of contradiction) for us to have any percep-tual experience of it at all. His transcendental framework is a way of understanding the limits of metaphysics and epistemology, which can then be used to pursue the traditional questions of philosophy. He concludes that all knowledge must begin from the *transcendental apperception*, that is, the introspective awareness of thoughts, much like Descartes' *Cogito*, except that Kant does not claim that we can know the "I" directly as a distinct entity. The "I" is not available to sensory perception and any reasoning about the "I" makes the "I" the object of our reasoning, and therefore, we cannot, by definition, examine it as it operates subjectively. So, to know any-thing, we must begin with basic perception.

Because Kant agrees with Locke and Hume that our perception is limited to the information directly present in our minds, Kant divides reality into two utterly distinct domains: the phenomenal, which is the world of sense experience and introspection; and the noumenal (**noo-men-ul**), the world that is not perceiv-able by us in any sense. We know of the noumenal world only negatively—e.g., there must be something outside our senses, at least space and time, since they are necessary for any experience at all. We know nothing about the noumenal beyond the conditions necessary for the existence of a phenomenal world. The phenomenal world is the world we know through experience; and from our experiences, Kant derives a handful of necessary conditions for knowing.

Kant's conclusions include claims like: The existence of time and space are necessary conditions for the possibility of even imagining objects and their relationships among one another; a law of universal causality is necessary for perceiving events in time; and all of the objects of our understanding are organized according to four categories: quantity, quality, relation, and modality.

After establishing the necessary conditions for knowledge through experience and reason, Kant considers the possibility of moral judgments—judgments about what we *ought* to do. An understanding of morality is a primary goal for Kant, because he thinks moral obligations are binding on us, and they are the sole motivation toward what is ultimately good for us. He concludes that there are three conditions necessary for moral judgment, and though they cannot be proven through any metaphysical or epistemic line of reasoning, they must be accepted speculatively as necessary conditions for moral judgment and action. These three conditions are: the existence of free will, an eternal soul, and God. Free will cannot be proven because the world as it is given in experience is mechanical and strictly law-governed (Kant was a strict Newtonian); nevertheless, it is necessary to believe we can act freely in order to accept that we are responsible for our actions. Similarly, the eternal soul cannot be proven because the "I" is elusive, and there is no evidence of how long we will survive into the future; nevertheless, the goal of morality is to achieve the highest end toward which reason is directed, namely a future perfect world, which can only be attained by living morally. And finally, the existence of God cannot be proven because God is not given in our senses or our reason, and arguments for and against God's existence are insufficient; nevertheless, morality requires a regulator (though not a giver) of a moral law, that is, a being to connect reason with good actions and direct good actions toward their proper end, namely, the highest good.

These conditions cannot be derived directly from reason, but are mere "transcendental" postulates, necessary for the possibility of reasoning about anything. Kant explains:

> It is necessary that our entire course of life be subordinated to moral maxims; but it would at the same time be impossible for this to happen if reason did not connect with the moral law, which is a mere idea, an efficient cause which determines for the conduct in accord with this law an outcome precisely corresponding to our highest ends, whether in this life or another life. Thus without God and a world that is now not visible to us but is hoped for, the majestic ideas of morality are, to be sure, objects of approbation and admiration, but not incentives for resolve and realization... (*Critique of Pure Reason*, "Doctrine of Method," Ch. II, Sec. II, A812–13).

Thus, free will, an eternal soul, and God form the necessary prerequisites for a complete understanding of morality. To be sure, Kant does not believe that God gives us the moral law; that is a product of reason alone. But he does think that the existence of God, who embodies the highest good, and a world in which to enjoy the rewards of a life well-lived are the primary motivations for acting morally.

Kant's writings are very technical and he coined so many new terms there are rumors that German and English Kant scholars consult one another's translations in order to better understand his view. Kant's views have been very influential, and aspects of it still permeate contemporary debates (including the *a priori/a posteriori* distinction). Nevertheless, critics of his arguments arose after several important developments in mathematics and science. Kant, like Newton, Locke, and many others, supposed that Euclidean Geometry was the only possible geometric system for describing space. Though other geometries may be logically possible, no others accurately represent how we perceive objects in space, neither can they compete with the precision with which Euclidean geometry allows us to predict the movements of objects in space. On this assumption, several claims seem to follow from Kant's metaphysics: space is necessarily Euclidean in structure, space is necessarily three-dimensional, and time is an entity distinct from space.

In the mid-1800s, after Kant's death, mathematicians discovered a host of non-Euclidean geometries (Riemannian, Lobachevskian, Circular, etc.). This wasn't an immediate problem for Kantians, since Kant allowed for the logical possibility of other geometries. But it opened the door for an interesting view of space. And soon, physics walked through that door.

In 1905, a young physicist named Albert Einstein developed a theory of physical motion more powerful than Newton's that combined space and time into a single entity (the Special Theory of Relativity). And in 1915, he developed a theory of gravity that substituted Riemannian Geometry for Euclidean (the General Theory of Relativity). These developments were difficult for Kantians, and for many, the General Theory of Relativity spelled the ultimate death knell for the remaining rationalist elements of Kant's philosophy. They felt they had no choice but to become empiricists. Geometry was now divided; one could study "pure geometry," which is analytic *a priori* in nature; or "applied geometry," which is synthetic *a posteriori*. Kant's synthetic *a priori* was all but eliminated.

Logical Positivism: Back to the Senses

The empiricism that evolved from this failure of Kantian rationalism is known as *Logical Positivism*, which was a revised version of a view held by nineteenth-century empiricist August Comte. Comte held a particularly strong version of empiricism that he called Positivism. Presupposing Comte's empiricism, the Logical Positivists, led by a group of mathematicians and scientists in Vienna, Austria (called the "Vienna Circle"), attempted to develop a mathematical language for expressing our perceptions of reality. The assumption of the group was that, even if science cannot be justified on philosophical grounds, it is still the most powerful tool we have (in their day, we had used it to build trains, cars, ships, and planes; now we use it to replace hearts, build skyscrapers, and play Guitar Hero). They worried that many of the problems of philosophy from the past few centuries were due, in large part, to ambiguities and vagueness in our language. The language that has been the most useful in science is mathematics (Einstein's Theory of Relativity, for example, started off as nothing but a simple equation). Therefore, as we develop theories about perception and reality, we should attempt to express them with as much mathematical precision as possible.

Their strategy was to believe nothing that is not: directly evident (mathematics and logic), evident to the senses (direct perception), or incorrigible (claims we can't *not* believe). This strategy led them to make advances in symbolic logic and philosophy of language. They developed interesting ways of expressing sensory claims. They hoped to derive all scientific claims using this technique. But many found that the project could not be completed. Rudolph Carnap, a member of the Vienna Circle, rewrote his theory twice in response to criticisms, and eventually abandoned it altogether. Other philosophers pointed out internal inconsistencies in the project. For instance, the very assumption—to believe nothing that is not directly evident, evident to the senses, or incorrigible—is not itself directly evident, evident to the senses, or incorrigible. By the mid-1950s, Logical Positivism was abandoned in all but spirit.

Edmund Gettier: A Change in the Curriculum

Epistemology was in a bit of a mess following the failure of Logical Positivism. Some tried to revive neo-Kantianism; some pushed to develop a weaker, internally consistent, version of Logical Positivism; others began developing what are now known as "naturalistic" theories of epistemology, many of which are still being developed and discussed today.

Despite the disagreement, most philosophers were still operating under the assumption that knowledge just means, "justified true belief." The question had just always been: How do we know when we're justified? But in 1963, a relatively unknown philosopher named Edmund Gettier, under pressure from the Administration of Wayne State University (where he was teaching at the time) to publish something, published a three-page paper that changed how philosophers approach epistemology. He challenged the very idea that knowledge is justified true belief.

Imagine that all the skeptical arguments were solved. Would having a justified true belief of some claim, *p*, be enough to conclude that a person knows that *p*? If a philosopher could show just one example of justified true belief that does not count as knowledge, then "justified true belief," while possibly necessary for knowledge, is not sufficient for it—we would need another condition.

Edmund Gettier constructs two examples of why justified true belief is not sufficient for knowledge. When philosophers offer an example to show that some claim is false, it is called a "counterexample." Gettier asks us to consider the following scenario:

> Suppose that Smith and Jones have applied for a certain job. And suppose that Smith has strong evidence for the following conjunctive proposition:
>
> (a) Jones is the man who will get the job, and Jones has ten coins in his pocket.
>
> Smith's evidence for (a) might be that the president of the company assured him that Jones would in the end be selected, and that he, Smith, had counted the coins in Jones's pocket ten minutes ago. Proposition (a) entails:
>
> (b) The man who will get the job has ten coins in his pocket.
>
> Let us suppose that Smith sees the entailment from (a) to (b), and accepts (b) on the grounds of (a), for which he has strong evidence. In this case, Smith is clearly justified in believing that (b) is true.
>
> But imagine, further, that unknown to Smith, he himself, not Jones, will get the job. And, also, unknown to Smith, he himself has ten coins in his pocket. Proposition (b) is then true, though proposition (a), from which Smith inferred (b), is false. In our example, then, all of the following are true: (*i*) (b) is true, (*ii*) Smith believes that (b) is true, and (*iii*) Smith is justified in believing that (a) is true. But it is equally clear that Smith does not *know* that (b) is true; for (b) is true in virtue of the number of coins in Smith's pocket, while Smith does not know how many coins are in Smith's pocket, and bases his belief in (b) on a count of the coins in Jones's pocket, whom he falsely believes to be the man who will get the job ("Is Justified True Belief Knowledge?" *Synthese*, 1963, proposition letters have been changed for clarity).

Gettier sets up a case where Smith infers some belief, (b), from a belief he has strong evidence for, (a). (b) happens to be true, but not for the reason expressed in (a). And, in fact, (a) is false. Now we have an interesting result: Smith believes (b), (b) is true, and Smith is justified in believing (b) [because (a) is evidence for (b), and Smith has good reasons for believing (a)]. So, Smith has a justified true belief that (b), but does not know (b). Why can't we say that he "knows" (b)? Because the fact that (b) is true is merely a matter of luck, and knowledge is not supposed to depend on luck. (b) is not true because (a) is true, but for some other reason. Since Smith believes (b) for the wrong reasons, his justified belief that (b) is true does not count as knowledge. If Gettier is right, this is a counterexample to the claim that a justified true belief is an instance of knowledge.

The Move to Externalism

What's an epistemologist to do? Gettier's counterexamples (and the dozens of others they inspired from other epistemologists) motivated an interesting shift in epistemology. Until this point in history, philosophers supposed that knowledge is something that, if you have it, you could be aware that you have it. The fact that you know something is something that would be available to your mind in some respect. This information would be among your beliefs; that is, it would be "internal" to your belief system. Therefore, the traditional approach to epistemology is called *Internalism*. Gettier's counterexamples forced philosophers to reconsider Internalism.

Most still agree that justified true beliefs are still necessary conditions for knowledge. However, they argue that some additional condition is also required, particularly, something that guarantees that the belief was formed in the "right" sort of way. How a belief is formed may not be directly available to your mind (e.g., knowledge may require a certain type of connection in your brain, or a certain type of relationship between the external world and your neural system—neither of which you can directly perceive). Because this condition would be "external" to your belief system (it would not be found among your beliefs), this approach to knowledge is called *Externalism*.

Some Externalists argue that a person's context determines whether their justified true beliefs are knowledge. For instance, if a person dreams that a political candidate wins an election, and it just so happens that, while the person is sleeping the candidate really wins the election, the sleeping person has a justified true belief that the candidate was elected, but the context in which the belief is formed is not appropriate for obtaining knowledge. This view is called *Contextualism*. Others argue that a justified true belief counts as knowledge only if it is produced by a belief-forming system or process that produces more true beliefs than false beliefs; that is, it is generally reliable. Though your belief-forming process may produce false beliefs from time to time, as long as it produces more true beliefs than false, the process itself is reliable, so this view is called *Process Reliabilism*. Some Process Reliabilists even drop the requirement that a belief must be justified internally to be an instance of knowing. As long as the process by which you obtain a belief is reliable, you do not need to know how you obtained it for it to be knowledge.

In the years following Gettier's article, various other approaches have arisen, and the field has expanded significantly. To explore this topic further, consider the suggested readings we have listed below.

QUIZ

1. Relations of ideas directly concern all the information provided by our senses.
   A. True
   B. False

2. Hume argues that even well-founded conclusions about what we can know through our sense experience are insufficient for justifying even _____ inferences about nature in general.
   A. probabilistic
   B. deductive
   C. valid
   D. natural

3. Hume shows that none of the three ways of establishing the truth of any claim can be sufficient for justifying our belief in induction. These three ways are:
   A. indirect sense perception, induction, and deduction.
   B. direct sense perception, induction, and deduction.
   C. indirect sense perception, deduction, and validity.
   D. deduction, validity, and direct sense perception.

4. What does Hume admit that he cannot live without in his daily life?
   A. justification
   B. belief-formation
   C. inferences
   D. induction

5. The traditional view is that we can know claims about mathematics and logic *a priori*, but nothing about the natural world.
   A. True
   B. False

6. Which of the following is a synthetic *a priori* claim?
   A. "5 + 7 = 12"
   B. "That wall is orange-colored."
   C. "All bachelors are unmarried males."
   D. "I think, therefore I am."

7. One experience contrary to a hypothesis (say, one black swan) can disprove a universal claim (that all swans are white). Therefore, scientists form hypotheses and then test them for disproof, a view known as
   A. inductivism.
   B. deductivism
   C. particularism
   D. presciptivism

8. **Kant divides the world into the "phenomenal" and the "neutral."**
   A. True
   B. False

9. **Their strategy was to believe nothing that is not: directly evident (mathematics and logic), evident to the senses (direct perception), or incorrigible (claims we can't *not* believe).**
   A. The logical extremists
   B. The logical positivists
   C. The logical prescriptivists
   D. None of the above

10. **Though your belief-forming process may produce false beliefs from time to time, as long as it produces more true beliefs than false, the process itself is reliable, so this view is called**
    A. Process Philosophy
    B. Process Epistemology
    C. Process Internalization
    D. Process Reliabilism

chapter 6

The Mind-Body Problem

CHAPTER OBJECTIVES

In this chapter, you'll learn about…

- The nature of mind
- Qualities and types of minds
- Ancient Western mythological conceptions of mind
- Descartes's argument for substance dualism
- The mind-body problem
- Contemporary theories of mind
- The staying power of consciousness

What Is a "Mind"?

Most of us assume that we have something called a "mind" and that this thing is as real as the brains in our heads. If you just take a moment to close your eyes, cover your ears, and think about what you were doing five minutes ago, you can probably imagine your "mind" engaged in the very thinking you were doing five minutes ago. But can you imagine your "mind" as a thing at all? Can you describe it? Does it have a shape, color, texture? If you can't sense it and can't focus on it, what is it?

In this chapter, we want to know: What is a "mind"? And it is important to realize that there is a lot of disagreement about the answer. There are those who think that:

minds do not exist at all and are a complete illusion produced by the brain.

minds are a spirit or soul that can be wholly detached from the body at death.

minds are the sort of thing that can be found, more or less, within the animal kingdom.

minds are the sort of thing only humans possess.

the entire universe is imbued with a cosmic, mind-like quality.

As good critical thinkers and philosophers, we don't want to simply assume that minds exist—even though it seems like they do—without offering arguments and evidence for their existence. So let's take a look at some of the most important views about mind found in the history of Western philosophy, and then you can form your own opinion about whether there are minds and, if so, what minds are like.

Qualities and Types of Mind

Most people in the world believe that minds exist, so, for a moment, let's assume they're right; minds exist. Let's try to roughly describe what they might mean by "mind." What kinds of qualities (features, properties, or characteristics) can be found in a mind? First off, it seems that most non-living things, that is, things that are not biological or human-made, do not have a mind. Planets, parks, pebbles, tornadoes, tables, and transistors, for example, don't seem to think about, or feel, anything, and so, are mind*less*.

If you think that a basic stimulus-response mechanism in a living thing is enough to qualify as a mind, then an amoeba has a mind, since it is able to respond to light and dark and move its little amoeba body accordingly. And most people think that an animal with a relatively complex nervous system (made up of a basic central nervous system and a basic peripheral nervous system) has a mind of some kind—and plenty of neuroscientists, psychologists, and other mind researchers think this, too. Vertebrates like bony fish, reptiles, amphibians, birds, and mammals would also meet this condition, and we often want to say that these animals have some type of mind, as well. Now, let's take these rough categories and try to organize them systematically.

The following are typical qualities of a mind (there are more divisions and distinctions, for sure, but we won't mention them here):

1. Perceptual Awareness (Perceptual Mind):

    the ability to recognize an object through some sense mechanism, as well as

    associate a stimulus with some memory,

    which requires a fairly small brain

*For example, fishes, lizards, and birds will move toward someone who is about to feed them because they seem to recognize, and/or remember, that it's feeding time.

2. Basic Reasoning (Reasoning Mind):

    the ability to perform a basic inference like "this is an animal that will eat me; therefore, I must get out of here," as well as

    the ability to solve a simple problem like using a stick to get at food just out of reach, which requires a bigger brain capable of storing more memories

*For example, cats, dogs, aardvarks, orangutans, and all other mammals will fight or flee, as well as share food, given a set of circumstances that requires them to do a basic "I need to think this through."

3. Consciousness (Conscious Mind):

    the ability to recognize oneself as an actor in some event,

    think about one as a self who is thinking,

    form beliefs about the past and future,

    imagine things that could not be directly experienced by the senses, and

    experience a range of emotions that are more than basic pleasures and pains,

    all of which require a brain with a fairly big frontal lobe

*For example, John believes that he could be President of the United States one day; Judy stands at the edge of the Grand Canyon, takes in the experience, and feels small in comparison; Mary devises a new hypothesis that explains another hypothesis of quantum mechanics; Joan reads this book and starts thinking about her own belief regarding the existence of a god; or Chris expresses the emotions of hope, then fear, then regret, when remembering a past event.

Notice that these qualities make it such that we can distinguish different types of mind: There is a type of mind that has perceptual awareness, call it Perceptual Mind; there is a type of mind that can engage in basic reasoning, call it Reasoning Mind; and there is Conscious Mind. If pressed, most people would say that if a thing has at least perceptual awareness, then that thing has a mind. Notice that an amoeba does not seem to have perceptual awareness (so, it probably does not have a mind), and the same goes for a lot of other species in the various biological kingdoms like bacteria, fungi, and plants. Insects could be considered a borderline case where they may or may not have mind. All vertebrates seem to have perceptual awareness, so they probably have minds. Pet fish, lizards, mice, cats, and dogs, as well as little children, easily recognize when it's feeding time, for example, so these animals all seem to be perceptually aware of what's going on around them.

TIP   *No one can say for sure, with absolute certainty that an animal does or does not have a mind because we can't get inside of their little heads, so to speak. Plus, most animals can't tell us what they're thinking, if they're thinking at all. However, we can and do perform experiments based upon human mental capacities, and this enables us to say that some individual of a species in the biological kingdom probably has a mind or not and, if it does have a mind, how advanced the mind might be.*

But, there seems to be a big difference between a marlin's mind, a mouse's mind, and a man's mind. Basic reasoning and consciousness are qualities of a mind, too, but most people would not say that a marlin, or even a mouse, is conscious the way a man is. The fact that humans are able to not only speak, write, theorize, and create works of art, but also solve quadratic equations, construct space shuttles, erect cities, and uphold civil and moral laws to benefit the weakest members of a society all seems to be evidence of the fact that humans are conscious.

Figure 6-1 represents ovals of the types of mind we have spoken about, as well as examples of vertebrates that exhibit that particular type of mind. There are things that we can all agree don't have minds—like chairs, cars, and planets—which are outside of the ovals. We have drawn the figure in such a way that the

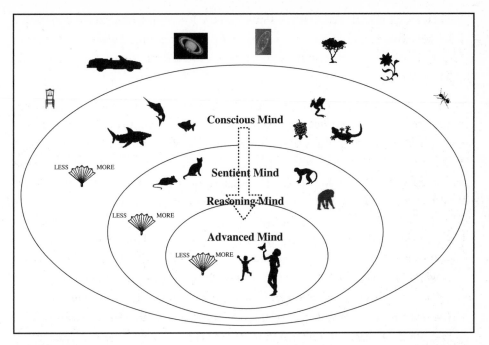

FIGURE 6-1 • Types of minds.

dotted arrow represents a kind of hierarchy of less complex mind to more complex mind, with:

Perceptual Mind being the least complex form of mind,

Reasoning Mind building upon, and being more complex than, Perceptual Mind, and

Conscious Mind building upon, and being more complex than both Perceptual Mind and Reasoning Mind.

So, for example, humans have Conscious, Reasoning, and Perceptual Mind, while cats have Reasoning and Perceptual Mind only, and fish have Perceptual Mind only. Also, the fan with the LESS and MORE indicate that there are *varying degrees* within the types of mind. For example:

it's probably the case that sharks have a less sophisticated form of Perceptual Mind than lizards, in general, have; but there are some species of amphibians that have a more sophisticated form of Perceptual Mind than most lizards have;

it's probably the case that mice have a less sophisticated form of Reasoning Mind than dogs, in general, have; but many species of primates have a more sophisticated form of Reasoning Mind than most dogs have;

it's seems to be the case that infants have a less sophisticated form of Conscious Mind than seven-year olds, in general, have; but university professors have a more sophisticated form of Conscious Mind than seven-year olds have.

Finally, there will be grey areas in our description of types of mind where it may be that: some non-vertebrate biological species have Perceptual Mind (perhaps some insects?); some reptiles have Reasoning Mind (maybe chameleons?); some primate species have Conscious Mind (like chimpanzees or gorillas?). New research from neurobiologists, psychologists, and others concerned with mind are documented almost daily in various respectable journals and books.

TIP *It's not the actual size and weight of a brain that is the indicator of a mind's complexity; it's the number of neural connections in the brain. An elephant's brain is bigger than a human's brain, but humans are obviously conscious and smarter than elephants. Why? A human brain has many, many more neural connections. The same goes for the now extinct* Homo neandertalensis *species with whom we co-existed for a time (Neanderthals went extinct some 30,000 years ago, or so). They apparently had bigger brains than we* Homo sapiens *do, but probably lacked the number of neural connections we have, and so we out-smarted, and thus, out-lived them.*

Ancient Western Mythological Conceptions of Mind

Now we have a way of talking about minds, but how could we really know whether we really have minds, or, if we do, what they are really like? For that, we have to turn to the tradition of philosophical arguments about minds.

Nowadays, most of us think that a brain of some kind (along with a functioning nervous system) is necessary for a mind to function, let alone exist. However, for a great deal of Western civilization's history most people thought that the qualities we now associate with mind referred to an entity that could exist without a brain, or any part of the body. Think of what people have referred to as a *soul* or *spirit* that lives on in some kind of an afterlife either as part of a heavenly place, on some other astral plane of existence, or as one with (or alongside with) some kind of god or gods. In many of these cases, the soul or spirit has some awareness, knowledge, and even emotional feelings independently of a body.

Ancient Egyptian, Greek, and Roman mythologists held similar views about the existence of a soul. For them, the soul was more like a *principle of life* or *the thing that caused a body to move.* Also, the soul was not limited to humans, but could be found in all kinds of living things; and even in some things we now consider natural but non-living, like tornadoes, rain, planets, and stars. Many myths from these cultures speak about the souls of cats, birds, dogs, and humans exiting the body once these living things died, just like when some cartoon image of a ghostly body (sometimes with wings) floats up to heaven when that cartoon character dies in cartoons.

Ancient Greek Conceptions of Mind

Most ancient oral and written traditions include myths, that is, stories that were accepted as valuable and that expressed sacred truths. Ancient Greek philosophers, such as Plato (428–348 BCE), believed that souls exist in much the same way as their predecessors in oral traditions believed, but the difference was that, being good philosophers, they actually gave *arguments* for the existence of a soul, rather than merely communicating stories. One basic argument they offered for the existence of the soul can be formulated like this:

1. Look around you: nothing moves by itself.
2. Anything you see always needs another thing to move it.
3. Natural things are no different from human-made things in that they need to be moved by something else.
4. Just like anything else, the bodies of living things can't move by themselves.
5. Thus, there has to be something else, other than the body, that moves a body, and this we can call a soul (or *psyche* in Greek). (This is a formulation of an argument found in Plato's dialogue, *Alcibiades*.)

> **BOX 6-1** CRITIQUE IT!
>
> This argument can be critiqued by pointing out that, while it may be true that nothing moves by itself, we need not jump to the conclusion that there must be something *wholly other than a part of the body* (or something material in the bodily realm), like some soul, that is responsible for moving the body. Bodies can be broken down into parts, and when seen in this way, we can imagine one part of the body moving another part (or parts) of the body. Plato and many of the Greek philosophers did not seem to consider this as a viable option.

For the majority of Greek philosophers, the soul exists as a source of movement for the body, and there were different types of souls, from lower vegetative souls found in plants, flowers, and trees, to higher animal souls found in rats, cats, and dogs. The highest type of soul was the rational soul found in human beings, and this soul had the reasoning and conscious capacities we normally associate with the human mind. In fact, the Greek word for soul, "psyche," is where the English word "psychology" comes from, and nowadays psychology is the science that studies the mind in all of its types and forms.

Given that the soul was believed to be wholly distinct from the body, acting as a source of motion for the body, many Greek philosophers—as well as the Egyptian, Greek, and Roman mythologists—subscribed to something in philosophy of mind known as **substance dualism**. "Duo" is the Greek word for "two," and according to substance dualism, a living thing is made up of two wholly distinct substances, (1) a soul/spirit/mind substance and (2) a body substance, each of which can exist apart from the other. Those who believe in the immortality of the soul, or reincarnation, are substance dualists because they think that the death of the body does not mean the death of the soul. The soul lives on after the death of the body. A lot of people on the planet have been, and continue to be, substance dualists of one sort or another.

Descartes's Argument for Substance Dualism

Substance dualism was the dominant view in the history of Western Philosophy from the Greeks through the Medieval period, when the Christian philosophy of Aurelius Augustine (354–430), St. Anselm of Canterbury (1033–1109), x St. Thomas Aquinas (1225–1274), and William of Ockham (1288–1348) dominated the field. During the Medieval period, Christian philosophers took Greek substance dualism and, in a sense, made the distinction between mind and body even more distinct. This was due largely to the Christian view that a Supreme God created the human soul from nothing, and infused a body with that soul during pregnancy. This soul was believed to be a unique entity, distinct from any material object, which lives on after the death of the body, eternally, either along with God in heaven, or absent from God in hell. By the time Aquinas was writing, most Christian philosophers believed that only humans had the kind of soul—complete with the conscious, emotive capacity we are calling an Advanced Mind—that could survive the death of the body. So, these philosophers would have answered NO to the question: Do all dogs go to heaven? (Poor pups!)

Following the Medieval period, skepticism once again reared its head, and this era became known as The Enlightenment (ironic, since it led many philosophers to become pessimistic of our ability to know). But, as we saw in Chapter 4, Descartes defeated many of these arguments and became the Father of the next era, known as Modernism. Descartes was also a Christian philosopher who was trying to make sense of the relationship between mind and body. In his writings, he noted many of the same qualities of mind we have spoken about already like perception, memory, reasoning, and consciousness. Like Plato, Augustine, and Thomas before him, Descartes was a substance dualist. He offered the following argument for substance dualism:

1. When I meditate on it carefully, I notice that my *body* seems extended in space and time, and can be broken down into spatio-temporal parts, though my *mind* does *not* seem extended in space and time.

2. Objects in space and time can be divided into a set of constituent parts. (For instance, my body is comprised of a head, hands, arms, legs, feet, hair, etc.), but my mind *cannot* be broken down into any component parts—it is completely "simple."

3. Therefore, my mind, which is not extended and has no constituent parts, is completely and utterly distinct from my body, which is extended and comprised of constituent parts.

BOX 6-2 CRITIQUE IT!

This argument begs the question, which means that the conclusion is already contained in the premises and, hence, does not really prove what it sets out to. Notice that the conclusion draws a distinction between body and mind; however, *the premises themselves already assume that distinction* since premise 1 claims that there is a distinct body while premise 2 claims that there is a distinct mind!

Since bodies and minds share no obvious properties, the mind must be made of something utterly distinct from whatever the body is made of. So, Descartes concludes that humans are made up of two substances, a thinking substance, or soul, and a bodily substance.

Concluding that the mind and body are so completely different leads to some interesting and difficult questions. For instance, if the soul substance is not in time and space, how does it interact with a brain at a moment in time at a point in space? And why do our soul substances hang around the bodies they do, why not switch out with others from time to time? Descartes admitted that he had few answers for these questions, though he thought his conclusions sound nonetheless. He did, however, suggest that the soul is connected by God to the brain at the pineal gland, a gland that is shared by both hemispheres of the brain.

BOX 6-3 CRITIQUE IT! Is Substance Dualism Scientifically Respectable?

Many philosophers and scientists argue that, from a scientific perspective, substance dualism is no longer a live option for at least one important reason (there are more): It seems that mental states are, at most, emergent or supervenient *properties* that are the result of brain states; they may not be reducible to brain states, but they are certainly dependent upon brain state processes. If there is any doubt about this, one need only peruse any textbook or journal devoted to the human brain's workings and read about the effects of brain damage upon the psychology of a person.

But is this a satisfactory response to the substance dualist? Dualists typically admit a great deal of dependence on brain states, and some even argue that the brain is a necessary condition of certain types of consciousness, namely, the experiences that are produced through our five senses by our brain. How might the substance dualist defend the existence of an immaterial mind in the face of this objection?

A Contemporary Argument for Property Dualism

Descartes' argument for substance dualism only reinforced this "commonsense" belief concerning the existence and nature of mind, a view that basically has been held by most people throughout the history of Western civilization. Now, many philosophical problems have been clarified or informed by scientific discoveries. This is especially so with philosophy of mind, given developments in psychology and neurobiology. Discoveries in neurobiology, psychology, and other sciences related to the brain have led many philosophers to replace substance dualism with a different version of dualism known as *property dualism*.

According to property dualism, a person is *one substance* that is made up of two wholly distinct sets of characteristics or properties: a set of immaterial mental properties (mental states) and a set of material bodily properties (the brain and neurobiological states). In this view, the mental states and brain states are distinct properties of some one person, similar to the way *roundness* and *blackness* are distinct properties found in the period at the end of this sentence. Just as we can distinguish the property of roundness from the property of blackness in some one period, so too, we can distinguish an immaterial mental property from a material bodily property in some one person.

However, just as the roundness and blackness of that particular period can exist only while that particular period exists, so too, according to property dualists, the mental and bodily properties of a person can exist only while that person is alive. So when we delete the period, the properties of roundness and blackness in that particular period cease to exist along with the period. Likewise, when a person dies, both that person's body and mind cease to exist (no brain, no mind). Such a view of mind in relation to body seems to be consistent with neurobiological and other scientific data, and is appealing to those who do not believe in the immortality or reincarnation of the soul.

> ### BOX 6-4 Descartes and the Pineal Gland
>
> Descartes argued that the mind interacted with the body through the pineal gland in the brain, probably because this gland is located near the center of the brain between the two hemispheres. At that point in Western history (1600s), most researchers thought that the brain was the basic material seat of the mind. Unfortunately, this is not a satisfactory solution to the problem of interaction since now the question becomes: How does the soul (a mental substance) interact with the pineal gland (*still* a bodily substance)!?! Descartes also drew the analogy that a soul is like a pilot and a body is like a ship which, although intending to respond to the problem of interaction, actually drove more of a wedge between body and mind since the pilot is a completely different thing from a ship, and can easily leave the ship.

The Mind-Body Problem (or Problems)

Despite the intuitiveness of substance and property dualism, many argue that these are just ways of avoiding the problem, and do not seriously address the nature of minds or mental states. The very nature of mental states (the fact that I can't be sure that you see "yellow" the way I see yellow; the fact that I can't feel your pains, etc.) raises what is known as the *mind-body problem*. The mind-body problem manifests itself in three ways:

1. From *both* the substance dualist and property dualist perspectives, there is what is known as the *problem of interaction*. How do these two things, which are so different from one another causally affect, and interact with, one another? How can an immaterial mind (either as mental *substance* or mental *property*) "move" my brain and nervous system to cause it to raise my arm, for example?

2. From the property dualist's perspective:

  a. How is it that these two distinct properties co-exist in the same substance? This is the *problem of co-existence*. No one ever sees, touches, hears, or senses in any way, a mind; yet anyone can see, touch, etc., a body or a brain.

  b. How is it that mental properties emerge from a set of physical properties in, say, the development of the brain or in evolutionary time, again, given that mind and brain are wholly distinct properties? This is the *problem of emergence*.

Material Monism

One way to address the mind-body problem is to reject both forms of dualism in favor of one version or another of what is known as *Materialism*, sometimes called *Material Monism* (whereas the Greek word for "two" is *duo*—hence, *dualism* in English—the Greek word for "one" is *mono*—hence, *monism* in English). The Scientific Revolution (ca. 1550–1750) in Western civilization inspired, among many intellectuals, the belief that all of reality is made up of spatio-temporal particles of matter of various sizes, moving around through forces such as gravity and other basic laws. All that exists is the "material" of nature, which motivates the name "Materialism." Materialists have been incredibly successful at explaining many different aspects of reality: think of all of the progress that has been made in the past 300 years concerning physics, chemistry, biology, medicine, space exploration, computer technology—all sciences concerned with matter in motion. Material monism comes in a variety of forms, including Behaviorism, Mind-Brain Identity Theory, and Functionalism. We will discuss each, in turn.

> ## BOX 6-5 A Word of Caution about Materialism
> Acceptance of material monism gets one out of the mind-body problems mentioned because there is just one thing—brain. So, no problem interacting, no problem co-existing, and no problem emerging! But, it has the obvious consequence of explaining away any mind or mental qualities as mere illusions, which most researchers and philosophers of mind nowadays find problematic.

Behaviorism

Given the success of physics, chemistry, and biology at explaining aspects of reality, by the beginning of the twentieth century, people doing research in mind started questioning whether the same kinds of scientific principles and methodologies found in these other successful disciplines could be applied to questions about mental states. One important feature of scientific study is the idea that in order to say that something exists, and in order for it to serve as a meaningful explanation of some phenomenon, that thing has to be *observed* by numerous people. So, physics, chemistry, and biology have been, and are, so good at explaining aspects of reality because these sciences only deal with data that can be observed, quantified, reported about, as well as tested and predicted. In the early years of the twentieth century, psychology was maturing into a respectable discipline, and many psychologists sought to explain the mind within the context of this scientific, and materialistic, perspective.

Psychologists such as Ivan Pavlov (1849–1936), John Watson (1878–1958) , and B.F. Skinner (1904–1990) thought that whether the mind really exists in some sense other than what can be directly observed, is irrelevant to science. We can say all that is relevant about mental states in terms of how they are manifested in the *behaviors* of animals, and/or in the dispositions of animals to *behave* in certain ways under certain circumstances. Thus, the science of *Psychological Behaviorism* was born. Using animal experimentation as the model, these behaviorists set up studies to show how various animals would react to certain stimuli presented to them. The reaction/action/behavior of the animal was then viewed as the animal's so-called "mind" and/or "mental states." Also, one could make predictions regarding behavior, too, since certain stimuli almost always yielded the same responses out of various animals. So, for example, a rat is given stimulus in the form of cheese in amaze, and its behavioral response (namely, it's "mind") is the movement of the rat through the maze to get to the cheese and eat it.

Behaviorists either ignored what was going on inside of an animal's mind, treating it like a "black box"—an insignificant aspect of the nature of minds—or they explained mind away altogether in favor of the behaviors exhibited by animals that could be observed by anyone. To be fair to the behaviorists, they were riding on the heels of the psychological views known as *Psychoanalysis* and *Analytic Psychology*, developed and defended by Sigmund Freud (1856–1939) and Carl Jung (1875–1961) who, from the behaviorists' perspective, seemed to put forward unverifiable, untestable, and unrepeatable accounts of the mind. Since these accounts left psychology powerless as a scientific project, they were discarded in favor of the empirically based Behaviorism.

Psychological Behaviorism must be distinguished from *Philosophical Behaviorism*. Philosophical Behaviorism is motivated by the same concerns that motivate psychological behaviorists, and they would agree on the psychology of the mind. However, philosophical behaviorists are concerned to know more than merely what explains our behavior; they also need a theory to explain how we talk about our mental life. We regularly express our "beliefs," "desires," "emotions." But if our minds are merely our behaviors, what do we mean when we talk about mental states?

Philosophical Behaviorists explain that, a set of statements about a person's mental states can be translated (fully, without loss of meaning) into statements about a person's actual or potential behavior. For example, the claim, "J believes it will rain" just means that J has some behavioral tendency that will manifest itself under certain conditions. So, it just means something like: "If J goes outside, he will take a raincoat; and if the tools are out, he will put them in; and, if the windows on his car are down, he will raise them; etc." The elegance of Philosophical Behaviorism is that we do not need to refer to anything beyond these behavioral dispositions to explain what it means for Jones to "believe" something or "hope" something or "feel" something.

A problem developed, however, as both Psychological and Philosophical Behaviorists attempted to explain more complex behavioral patterns. For instance, isn't it possible for a person to believe something that has no effect on his behavior whatsoever? Can't someone be angry or sad, for example, and give no bodily sign of it? Of course someone could.

As an example, imagine that someone, J, knew nothing about Einstein's Theory of Special Relativity, and now someone explains that the theory entails that as objects move close to the speed of light, they contract. J then forms the belief that objects moving near the speed of light contract. Does this need to have any effect on his behavior? You might say that he would answer differently now if he were asked a question about Special Relativity, than before he was told about. But that is true only if he is comfortable talking about his beliefs. It may be that he would answer the same. But would we then say that he doesn't *actually* have the belief? That seems absurd.

Philosopher of Language, Noam Chomsky (b. 1928), offers a different sort of argument against Philosophical Behaviorism. Chomsky argues that all languages have a few essential components, and that these are innate in our psychology— they are not conditioned by our environments. His main arguments for this view depend on the empirical evidence of how fast children learn their native language and their rich knowledge of its structures, despite very little exposure to the language. Therefore, what distinguishes humans from animals is an innate awareness of these essential grammatical structures—a "universal grammar."

But a universal grammar is something that, in principle, could not be explained in terms of behavioral dispositions alone.

Behaviorism in psychology soon encountered insuperable objections as well. Psychologist Edward Tolman (1886–1959), experimenting with Behaviorism, had mice learn a maze by walking it in order to find a piece of cheese. If Psychological Behaviorism is right, a mouse's "knowledge" of the maze would merely be a product of the conditioning of certain muscles, so that, if he filled the maze with water, the mouse should have to "learn" the maze all over. But that's not what happened. In a maze filled with water, mice swam straight to the cheese. Moreover, in another experiment, Tolman didn't even let the mice walk the maze; he put them in tiny wheel barrows and drove them through the maze. As it turns out, they could find the cheese again as quickly as any that had learned the maze by walking it. The conclusion was devastating to Psychological Behaviorism: mental states are not merely a product of behavioral conditioning.

BOX 6-6 CRITIQUE IT!

The behaviorist's claim that the mind is either a black box or can be understood as behaviors has been critiqued and rejected for the main reasons that:

the black box of the mind can *and should!* be opened, examined, and explained, as many cognitive psychologists, neuroscientists, cognitive scientists, and others think

the mind seems to be completely distinct from behavior, acting as the *motivator or cause* for behavior, as when:

John *desires* to eat ice cream, *then* behaves accordingly by eating ice cream

Sally *believes* that the car coming toward her will harm her, *then* behaves accordingly by moving out of the way of the car

Chris *thinks* that abortion is immoral, *then* behaves accordingly by not voting for Senator Smith

most significantly, mental states like perception, thinking, beliefs, emotions, and especially conscious awareness, are things that are believed by many to be real and do exist somehow.

Mind-Brain Identity Theory

The failure of Behaviorism led many materialists to a more extreme view of the mind. Rather than ignoring it in favor of behaviors, they simply identified mental states with brain processes. The idea is that, mental states should not be regarded as a black box; rather, we should regard them simply as brain states. If we are studying brain states, we are studying no less than mental states themselves, a view known as *Mind-Brain Identity Theory*.

The motivation for Mind-Brain Identity is simple: just as we discovered that lightning is nothing more than electrical discharge and water is nothing more than H_2O, we will eventually discover that "seeing yellow" is nothing more than a complicated brain state. Science has rid us of our tendency to treat lightning as the whim of the gods, and the tendency to believe that biodiversity requires a supernatural designer; therefore, if we are patient enough, it will dispel our tendency to believe there are non-physical minds above and beyond our brains.

A popular version of Mind-Brain Identity theory is called *Eliminative Materialism*. Eliminative materialists believe that any mental quality can be identified with an underlying neurobiological process and, hence, can be wholly identified with that neurobiological process, so that even our perception of mental states is illusory. They also argue that any mind-talk is really like myths or folklore that should just be removed from our thinking and vocabulary altogether. Just as mythological gods, witchcraft, and the geocentric view of the universe were found to be nonsense, so too, the neurobiological sciences will progress to the point where we can stop using mental terms like *consciousness*, *belief*, *perception*, *opinion*, and *pain*.

Why believe Eliminative Materialism? Defenders often point to the fact that we can already explain a lot of "what's really going on" with the brain in purely neurobiological terms. For example, when I say, "I am experiencing a headache" that can be replaced with a kind of sterile, hyper-technical neurobiological speak, such as "Group of neural cells X in area Y of my frontal lobe is being pressed upon by vein Z given that vein Z is pumping blood at an accelerated rate." So, the eliminative materialist bets that the *experience of pain*, the *imagined idea* of the Garden of Eden, the *belief that* Global Warming is occurring—all mental things—will be eliminated, explained away by appealing to "what's really going on," namely, a series of material, neurobiological events associated with the brain and nervous system.

In addition, defenders also highlight how easy it is to alter a person's mind:

1. By damaging a part of the brain or nervous system, a person's mind is altered, and sometimes permanently. For example, without the normal functioning of the prefrontal cortex, individuals are not able to make plans, nor are they able to carry out the behavior necessary to fulfill those

plans. Also, damage to the prefrontal cortex causes a person to be unable to store short-term memories. Further, damage to the limbic system can cause certain autisms and other emotional dysfunctions.

2. By applying neuro-pharmacological drugs, we can alter a person's mental states. For example, doctors prescribe selective serotonin reuptake inhibitors to regulate depressive and anxious moods. Once they've taken their meds John no longer feels like he wants to kill himself, and Sally no longer experiences anxiety and wants to wash her hands 250 times a day.

So, a basic argument for mind-brain identity goes like this:

1. It seems as if we live in a physical world of spatio-temporal entities whereby physical entities affect other physical entities.

2. Neurobiological processes and the actions of animals (including humans) are not exempt from these relationships, as they affect, and are affected by, spatio-temporal entities. For example, material neurobiological parts and processes move material physiological parts and processes to move my material arm to move the material cup up to my material lips to take a drink of material water.

3. Countless experiments and observations suggest that a mental quality is dependent upon an underlying neurobiological process. In other words: adjustment to brain means adjustment to mind, damage to brain means damage to mind, as well as *no brain means no mind*.

4. Add to this the problem of interaction between a mental quality and a neurobiological process (as with substance *or* property dualism) whereby there just is no satisfactory explanation for how it is that the mind affects the brain and vice-versa.

5. Hence, the mind is really identical to the brain, and the various manifestations of mental qualities (thoughts, beliefs, emotions, etc.) are really brain processes of one kind or another.

BOX 6-7 CRITIQUE IT!

1. This argument's premises don't guarantee the truth of the conclusion, but merely offer grounds for the likelihood of the conclusion being true. That said, it could be argued that there is an underlying *presumption* that the mind does not exist found in the premises themselves, which would make the argument be a question-begging one.

2. Another way to critique this argument has to do with showing what contradictions or other logical problems arise if we accept the truth of the conclusion. So, **if mind is identical to brain, then**:

    For example, my mind's thinking the thought of the color green should have a material, localizable green counterpart found somewhere in the brain. However, my thought of green does not have such a counterpart (how could it? and where exactly would it be located?), so mind is not identical to brain.

    My thoughts, beliefs, emotions, perceptions and other mental qualities are all just illusions. However, it seems like those things are more real to me than anything else, including my own body, which I can imagine being completely different than how it is now. These mental qualities are mine, I own them, and it seems I would not exist without them; so, how could they be illusions?

    Mental qualities should be isomorphic, and exactly localizable, with certain neurobiological processes. However, research shows that (1) not only do animals have slightly different neuro-physiological compositions, but also that (2) various neurobiological processes are flexible enough to perform other functions as when, say, a part of the nervous system is damaged. So, such mental qualities are not isomorphic, and exactly localizable, with certain neurobiological processes.

Like Philosophical Behaviorism, Identity Theory faces serious criticisms. In order for two objects to be identical, they must share all the same properties. For example, Ball A is identical with Ball B if and only if Ball A and Ball B are the same size, shape, weight, density, in the same location, related to other objects in exactly the same way, etc. But once we see that they agree in all these properties, we recognize that we're not talking about two different things, but only one self-identical thing. Now, consider mental states. We imagine all of our readers familiar with the United States agree that Denver is the capital of Colorado. If we agree, it would seem we share the *same belief*. If so, what are we "sharing"? We *are not* sharing the same brain state. But presumably we are sharing something. But if the brain state has a different property than the mental state, namely, something that can be shared, they are not identical.

Similarly, the neural state corresponding to my thought "this mug is red" is 2 inches inside my right ear. But is the *thought* 2 inches inside my right ear? It seems unlikely that thoughts have spatio-temporal locations. And if they cannot, since brain states do have spatio-temporal locations, brain states are not identical with mental states.

Functionalism

Given the critiques mentioned above (and many others not mentioned), toward the latter part of the twentieth century many researchers doing work in philosophy of mind not only found straight-forward substance dualism unreasonable, but also found versions of Behaviorism and Mind-Brain Identity Theory unsatisfying. These researchers thought that, although the methodology associated with science and materialism is correct in general, and the mind depends upon the brain (rejection of substance dualism), mental states also seem to be more than mere behavior and not wholly identical to brain processes. Of course, there are those who still hold to versions of all of these positions, but many philosophers of mind reject them.

By the 1950s, serious work was being done in computational systems and the development of the first electronic computers. From this work, researchers started thinking that the mind was like computational processing, while the brain was like the computational hardware which enabled the computational processing to take place. Mental states might be a function of brain states in the same way that a software program is a function of some piece of hardware. This view is known as *Functionalism*.

According to Functionalism, the mind is the totality of functioning mental processes at work in some system, along with the behavioral outputs indicating the results of the mental processing. We say "some system" because of the fact that a functionalist thinks that any system set up the correct way to be the bearer of a mind could, in fact, have a mind. As long as all of the correct parts and processes necessary for producing mental qualities are there, and those parts and processes are functioning correctly to produce mental qualities, then that thing has a mind.

This idea is kind of an example of where "science fiction meets science fact" or vice-versa. Because of this perspective, the science of *artificial intelligence* (AI) was born, where a big part of what AI researchers do is: set up computational systems; run these systems through various tests; document the results of their performance on the tests; and then use the data from these tests to better understand the mind. If in fact minds are like computer programs running on the hardware of the brain, then we should be able to develop highly complicated computer systems that will mimic human minds.

According to the functionalist of the AI variety, not only is it the case that a correctly functioning brain has a mind (of different varieties, of course, depending upon how sophisticated the functioning brain is), but it's also the case that a robot with a computational system set up just like a correctly functioning brain *could have a mind*. And further, we could encounter alien life-like forms that have minds, too, without brains per se, but having something (call it plasma) that functions just like a brain. Whereas the behaviorist's dictum could be "The mind is behavior" and the mind-brain identity theorist's dictum could be "The mind is the brain," the functionalist's dictum could easily be "The mind is how the brain (or something set up just like the brain) functions." Today, the idea that the mind is like a computer is commonplace, and we even use computational terms like *memory, storage, information, input, output, processing*, and others when we're talking about mental qualities and processes (see, we just used this computationally-based word, *processes*, again to talk about mental states!).

BOX 6-8 CRITIQUE IT!

The functionalist's position is an intriguing one, especially when we think about the possibility of artificially intelligent androids, robots, and droids existing sometime in the future, like we read and see in science fiction stories. However, many have found it hard to believe the idea that, if parts and processes are set up the correct way to function so as to bring about a mind, then a mind will somehow exist as part of, or emerge from, those functioning parts and processes. For example, there is obviously a complex system of communication that takes place in the brain and nervous system. So, what if we set up some kind of complex network of communicators, like a bunch of people with walkie-talkies sending messages back and forth to one another, similar to what takes place in the nervous system: we could hardly say that this entire complex network of communicators has a collective mind, could we? The same goes for the complex silicon and metal connections and complex interactions that take place with computational systems: we could hardly say that this computational system has a mind, could we? Thus, many have argued that parts and processes that functioning correctly is not enough to generate a mind.

We have to be careful drawing too closely the parallel between minds and computers. Reasoners who aren't cautious may be tempted to reason as follows:

1. Minds compute things, reason about things, have memory, and can learn.
2. Computers compute things, reason about things, have memory, and can learn, too.
3. Thus, all minds are computers.

BOX 6-9 CRITIQUE IT!

This argument, however, is an example of fallacy. A fallacy is an error in reasoning and usually has to do with someone inappropriately drawing a conclusion from premises when, in fact, the conclusion does not follow from the premises. So, in this case "the mind is a computer" does not follow from the premises, as the following analogous example argument shows:

Premise 1: Airplanes have wings and can fly.

Premise 2: Birds have wings and can fly.

Conclusion: So, airplanes are birds.

Just because airplanes and birds do the same kinds of things does not mean that *they are the same thing*; the same goes for minds and computers. At best, we could argue that minds *are like* computers, which many people take to be a reasonable conclusion.

It is interesting to note that people doing work in AI are now trying to set up computational systems to mimic the complex network of connections in the brain and nervous system. It may be that the only kind of system that could generate a mind is, in fact, the brain, or one that would have to *exactly, precisely, completely* imitate the brain, with all of its complex parts, processes, and evolutionary history.

Unfortunately, AI has not made the advancements that were so hopeful early on. In fact, there seem to be a number of mental states that we cannot figure out how to replicate. We have come a long way, writing computer programs that write orchestras at least as pleasing as any written by humans. The problem is that, of the orchestras written by the programs, the pleasing ones are a disappointing few. Similarly, there is no way to know (and no evidence that) the program enjoys the music it writes...or even understands it.

Still Struggling

At the beginning of this chapter we set out to answer the question: "What is a 'mind'?" And we noted that there is a vast amount of disagreement about what a mind is. Now that you've encountered a few theories, you may be feeling overwhelmed. But before you give up, consider how far we've come. We now know that, even if we are not logically identical with a brain, having a brain of a certain sort is physically necessary for having the sorts of perceptions we have (the experiences tastes, colors, scents, sounds, and textures we have; the memories we have; the reasons we understand). We also know that, if there is any non-physical component to our minds, they stand in a unique relationship to this brain of ours. And finally, we know that "consciousness" is so unique that our best neurology hasn't caught up with it.

What does all this indicate? It indicates that questions about the precise nature of the mind are still open to a large degree and that philosophers and neuroscientists need to spend a lot of time in the same bars after work.

The Staying Power of Consciousness

Despite all the advancements in neurobiology, there remain some aspects of the mind that seem in principle impossible to explain from a Materialist perspective. Frank Jackson asks us to imagine a color scientist who is raised in a completely black and white world. She knows everything about the physics of color and the physiology of color perception, but she has never seen color. He then asks, does she know all there is about color? It would seem not. If you were to escort here into the multi-colored world, it seems that she would discover something that no color science could capture—what it is like to *see colors*. Similarly, Thomas Nagel argues that, even if we knew all there is to know about the physiology of bats, we still would not know *what it is like to be* a bat.

Something about mental states resists scientific explanation. Whatever it is, it has led some philosophers back to dualism; it has led other to suspend judgment until more data is in. But all can agree that, in spite all of our theories, minds are things that still elude explanation, leaving us still in the throes of the mind-body problem. Recall that one's *subjective experience* of self, others, events, etc. is one of the features of the Conscious Mind. And it seems almost magical or miraculous that there is this whole other realm of conscious experience from which so many imaginary thoughts, complex ideas, and intricate conceptualizations have been generated and then acted upon throughout human history in the forms of art, technology, philosophizing, creative problem solving, and the like. Yet, conscious experience evolved along with the brain of *Homo sapiens*, and—gosh darn it!—there must be an acceptable explanation for how this happened and how conscious experience interacts with the brain. It is no wonder, then, that many still find the mind-body problem so… wonderful.

QUIZ

1. Which of the following probably does not have a mind?
   A. A bony fish
   B. A reptile
   C. An amoeba
   D. A bird

2. The ability to recognize an object through some sense mechanism is a capacity of:
   A. Perceptual mind
   B. Conscious mind
   C. Reasoning mind
   D. None of the above

3. Ancient Egyptian, Greek, and Roman mythologists held similar views about the existence of a soul.
   A. True
   B. False

4. While it may be true that nothing moves by itself, we need not jump to the conclusion that there must be something _____ that is responsible for moving the body.
   A. we can't see, but would like to when we die
   B. wholly other than a part of the body
   C. epistemologically-motivated
   D. metaphysically-obfuscated

5. Would a medieval Christian philosopher have believed that dogs went to heaven?
   A. Yes
   B. No

6. According to _____ , a person is one substance that is made up of two wholly distinct sets of characteristics or properties.
   A. property dualism
   B. substance dualism
   C. functionalism
   D. mind-brain identity theory

7. Behaviorists either ignored what was going on inside of an animal's mind, treating it like _____, or they explained mind away altogether in favor of the behaviors exhibited by animals that could be observed by anyone.
   A. a black box
   B. an invalid instrument

C. an evil thing

D. a necessary evil

8. **For the functionalist, mind is to software as brain is to**

A. Software programming.

B. Software program.

C. Hardware.

D. None of the above.

9. **What conclusion can be drawn from these two premises? Premise 1: Airplanes have wings and can fly. Premise 2: Birds have wings and can fly.**

A. Airplanes are birds

B. Birds are airplanes

C. Both of the above

D. Neither of the above

10. **Thomas Nagel argues that, even if we knew all there is to know about the physiology of bats, we still would not know _____.**

A. the bat's personality

B. what it is like to be human

C. what it is like to be a bat

D. what it is about the bat's personality that we like

chapter 7

Personhood and Personal Identity Over Time

What does it mean for you to be "you"? Are you a "person"? Are you the same person now as three years ago? If you lost your leg, wouldn't you still be "you"? A difficult metaphysical question concerns how we would answer such questions. In this chapter we will explain the motivation for these questions and discuss some of the traditional attempts at answering them.

CHAPTER OBJECTIVES

In this chapter, you'll learn about...

- Criteria for personhood
- Personhood and morality
- Personal identity over time
- Conscious minds in bodies
- Memories and identity
- Necessary and sufficient conditions

Are You a Person?

We presume that you, the reader, would consider yourself a person, right? In fact, a lot of people believe that every human being who comes into existence at the moment of conception is a person. However, especially since the early 1970s with the U.S. Supreme Court decision regarding abortion (*Roe v. Wade*, 1973), this view has been challenged, and a host of alternative perspectives have arisen. There are those who argue that:

Only people who are fully conscious and can take full responsibility for their actions are persons.

Human embryos, unborn children, and even young children are not persons.

People are not born persons, but normally developing people become persons later in life when they are fully conscious and can take full responsibility for their actions.

You can be a person one day, then the next day, after a tragic accident where you are in a coma or persistent vegetative state, you are no longer a person (or maybe you become a different person).

Who you are as a person can change over time, depending on what happens to your body or your mind.

In this chapter we'll take a look at some views about what it means to be a person and touch on the topic of personal identity over time. You'll then be able to form your own opinion about what it means to be a person, as well as what part of you stays the same as you move through time, growing and aging.

Criteria for Personhood

We will begin by laying out some generally recognized criteria for personhood, and then investigate each one. The following is by no means conclusive, but provides a good starting place. A person is a being who has the capacity to:

1. reason or think logically;
2. believe, intend, desire, emote;
3. understand and/or speak a language;
4. enter into complex social relationships;
5. act autonomously, fully responsible for actions committed or omitted.

Below we'll investigate each of these criteria, but before going any further, a couple of clarifications need to be made. First, (5) presupposes to a great extent that a being already possesses (1)–(4). At first blush, for example, we could say that the current President of the U.S.—Barack Obama—is a person in the sense of (5), but he also has many relationships with other persons, he uses language, has mental states, and reasons (though some question the latter). This presupposition will become evident as we go through the chapter. Therefore, some of these conditions may depend on others.

Second, notice that we used the phrase *being who has the capacity to*, rather than *being who does*. This is because we want to make sure that the definition of person is neither too narrow, nor too broad. If we said that persons were beings who exhibit traits (1)–(5) in a kind of absolute sense *at all times*, then our definition of personhood would be *too narrow*, that is, it would leave out beings who are obviously persons. For example, you probably qualify as a person because traits (1)–(5) are applicable to you. However, you are probably still a person when you are in a deep sleep yet not reasoning or intending, etc. So if we say that, as a person, you have traits (1)–(5) in the sense of *actively engaging in them at all times*, then when you are asleep you are not a person, and this seems absurd. However, if we say that you *have the capacity for* (1)–(5), then you can be a person, even when you are sleeping or in a coma, and you may be a person before you're old enough to reason, and not at that moment engaging in any of (1)–(5).

BOX 7-1 CRITIQUE IT!

Someone might object that we could never know if a being could meet the criteria for personhood because all we can observe are the outward actions of another being, and that it is not possible to draw any conclusions about the *internal cognitive workings* of a being from its/her/his *external actions*. It seems that all we can say we know for sure are *our own cognitive capacities*, namely, our own thoughts, our own beliefs, our own feelings, our own fears, our own perspectives, etc. We can never "get inside someone else's head," and so *I* can only infer that if some other being appears to be relative similar to us in bodily appearance, walk, talk, etc., then that being must have a similar set of cognitive capacities as *I* do.

To this, one could respond that while it is true that we can never draw any conclusions with *certainty* about anyone's cognitive capacities just from observing their behavior, we can feel justified in drawing *probabilistic* conclusions anyway. In fact, if we did not feel justified in drawing an inference about someone's internal cognitive capacities from their external behavior, then the sciences of psychology, sociology, and neurology would cease to exist altogether. Further, a neurosurgeon would never feel justified in operating on a part of someone's brain for the purposes of say, adjusting that person's mood swings, because he would think "Well, I can't know *for sure* what's going on inside that person's brain, so I don't know *for sure* if this operation is going to work." But obviously and thankfully, neurosurgeons do not think this way.

Also, the word *capacity* is used to prevent the definition of personhood from being *too broad*, that is, we don't want to include things that obviously aren't persons. Most philosophers would agree that being a person is not a matter of having a certain type of body. For instance, something is not a person just because it has human DNA. A being can be considered as a member of the human race, but still not be considered a person. Let's define a human being as: *a being that has all the genetic traits of* Homo sapiens, *as biologists would understand these traits.*

Now, it seems possible that a being be human in this sense, but lacks some essential ingredient of personhood. For instance, a dead human still has human DNA, but is no longer a person. Similarly, a severed finger or limb has human DNA, but is not a person. It is also possible that shortly after a human female egg is fertilized but before cell growth begins, an organism has human DNA, but is not a person.

I (as a Person) Ain't Got No Body... but I Got a Brain

Although someone might argue that very small children and the severely mentally handicapped are not persons, few would say that the *physically* handicapped are not persons. This raises the interesting matter of whether a body is even necessary as a condition for personhood. Note that, in the criteria for personhood, there is no direct mention of a physical body. Important implications can be drawn from this omission.

First, what it means to be a person seems not to be directly tied to an intact bodily existence. Take someone like the famous scientist Stephen Hawking. Here is someone whose body is ravaged by a neuro-muscular disease, and who needs machines in order to communicate. Yet, we would still consider him a person because, despite his bodily limitations, he fulfills criteria (1)–(5). He does this because his *brain* is still functioning the right way. Even the character Johnny, from the famous story and film *Johnny Get Your Gun*, whose arms and legs have been blown off in battle is still a person. Even though Johnny's bodily functions are impaired (or no longer there because he is basically a torso and a head lying in a hospital bed), his cognitive capacities are still functioning. Like Hawking, his cognitive capacities are there *because his brain is still intact*.

So, one might argue that (a) *cognitive capacities* are the *real* capacities to look to in a being when trying to discern whether a being qualifies as a person, and (b) the brain, or possibly something that functions like the brain, is a necessary condition of this cognitive capacity.

My Friend is a Brain in a Jar

Doctor Who is a cheesy British science fiction television series that has been running since the mid-1960s. There's an episode from the 1970s called "The Brain of Morbius" where an evil character named Morbius is kept as a brain in a jar of liquid by his henchmen. Morbius communicates with his henchmen through an electronic system connected to the jar. Eventually, the henchmen connect the jar that has Morbius's brain to the head of a hideous monster with a giant claw, and Doctor Who engages him in physical and mental battle. The point is made clearly that Morbius's identity as a deceptive, evil mastermind is linked directly to his brain and, given the reality of people existing like Johnny from *Johnny Get Your Gun* mentioned above who are missing all their limbs, science fiction is not far from science fact here. It really seems possible to be friends with a *person* who is a brain in a jar.

A Person...but not a Human?

Further, if having a body is a contingent feature of personhood and cognitive capacities are the necessary conditions, then it seems plausible that some robots (like C-3PO and R2-D2 from *Star Wars*), androids (think of Data from *Star Trek: The Next Generation*), and aliens (those little Martians who try to take over the Earth in the Tim Burton film, *Mars Attacks*) could be persons, provided their cognitive capacities play the role that persons' cognitive capacities do. We tend to think that persons must be biologically-based entities who breathe air, metabolize carbohydrates, and take in water for nourishment. However, consider the following thought experiment.

Right now it is possible to simulate various biological parts of bodies artificially; there are artificial hearts that pump blood, artificial kidneys that filter urine, and even artificial eyes that process visual stimuli. Suppose that a scientist develops an artificial occipital lobe out of silicon and metal, and implants it into the brain of an adult female human being. The artificial occipital lobe performs the same functions that a natural occipital lobe performs, namely, processing visual information from the environment. So, with the artificial occipital lobe she could do the same thing that she could do with her natural occipital lobe; she could see the world around her. So far, she still seems like a person.

Now imagine that the scientist also develops artificial silicon and metallic parts of the brain responsible for memory, and implants these into our female subject's brain. Again, she can store and recall memories with the artificial parts of the brain in the same way she could with her natural parts. Even now, she still seems like a person.

And imagine further that the scientist develops an artificial silicon and metal *brain in its entirety*, and implants it into our female subject. With this artificial brain, she can do all of the same things she did before her transplant; she lives, loves, lies, etc., and meets all of the criteria for personhood. Would she still be a *person*, even given that her brain is *robotic*? It is difficult for us to say no.

Now, say the scientist can simulate all parts of her body with silicon and metal, and replaces her biological body with a robotic body. She now is *fully* a robotic being in body and brain with all of the same hopes, fears, responsibilities, loyalties, etc., as any other human being who is a person. Would she (or should we say it?) actually be a person? How about replacing all of this stuff with an alien plasmatic substance that functions like the brain? You get the picture.

The point is that it doesn't seem that a thing *necessarily* needs to have a brain in order to think, believe, feel, experience, etc., if such cognitive capacities can be *simulated* by other means. So, it seems that a functioning brain, or something that *functions like* a functioning brain, with all of the cognitive capacities associated with such functioning, becomes what is significant in determining whether something qualifies as a person.

Still Struggling

How do we go about defining a term like "person"? To get a clear definition of an abstract concept like "person" or "mind," philosophers often begin with what is commonly taken to be a paradigm example of the concept. So, to ask what counts as a "person," we start with adult humans rather than mosquitoes or computers. We have to be careful, here, because we cannot claim from the beginning that mosquitoes and computers are *not* persons—we don't know yet. We simply begin from a paradigm example and then try to figure out what is essential to calling that thing a person. For example, we've seen that having a hand or a leg is not essential to being a person—you can lose either without ceasing to be you. We've also seen that having biological organs is not essential—we could replace them with prosthetics without your ceasing to exist. What's left? As we will see below, one of the most plausible definitions of what makes a person a person is the capacity for reason.

Criterion 1: The Capacity for Reason

We can now begin to investigate the five criteria for personhood a little more closely. The first criterion has to do with the capacity for reason or rationality. Rationality involves a variety of traits, including:

(a) calculating;

(b) drawing new conclusions or inferences from old information;

(c) forming associations between present stimuli and stored memories;

(d) problem solving.

If what is meant by rationality is (a) calculation, then it seems that a simple computer—like a calculator—has this ability. But no one thinks that a computer is a person like you or Barack Obama! Computer programs are pretty sophisticated these days and are able (b) to draw inferences and learn new things, too. For example, there are computer programs that are able to figure out the past tense of a word based upon learning other words and the contexts in which a word is supposed to be used. Similarly, programs like Genius from iTunes attempts to select songs from a particular genre or artist (well, some of these programs are better than others, but you get the idea). This conclusion might be arrived at by a process of reasoning that looks something like this:

1. The past tense of "turn" is "turned," the past tense of "own" is "owned," and the past tense of "look" is "looked."

2. <u>A new word has been input into my system, and it is the word "learn."</u>

3. Thus, the past tense of "learn" is "learned."

This is obviously too simplistic, and the number of exceptions to this rule would need to be exhaustively formulated. Nevertheless, we just taught a piece of plastic and silicon to add an "ed" to the end of a word in the right context. That's pretty amazing for a machine! However, even with this ability, we would not say that a computer is a person like you or Barack Obama.

Many animal species seem to be able to (c) form associations between present stimuli and stored memories, as when fishes, lizards, birds, and basically every kind of mammal at the zoo will move toward someone who is about to feed them because they seem to recognize, and/or remember, that it's feeding time. But, again, we would not say that these animal species are *persons*, would we?

More recently, we've learned that a lot of primate species are able (d) to solve fairly sophisticated problems, which seems to demonstrate some degree of rationality. For example, chimps can stack boxes or use sticks to get at food high up in the air, gorillas use rocks to crack open nuts, and orangutans can use leaves like spoons to carry and drink water. However, besides the fact that the number of possible solution routes is limited in these animals in comparison to a developing child, we would still not want to say that these primates are persons. At least, not *persons* like you or Barack Obama.

What all of this seems to show is that the capacity for reason—namely, intelligence understood as (a)–(d) above—is simply not enough for personhood. Another way to say this is that the capacity for reason may be *necessary* for personhood, but certainly not *sufficient* for personhood: Rationality is needed, but it's not all that suffices.

BOX 7-2   A Puzzle for Your Reason-Loving Mind

Test your intelligence and problem solving skills with this logic problem: The Smith family is getting ready to have a baby. The Smiths' are made up of a mom, a dad, a daughter, an older son, and a younger son. From the information below, determine the girl's name and the boy's name suggested by each family member.

> Neither of the women in the family suggested a girl's name beginning with the letter "D"; however, the woman who doesn't think that a girl should be named Linda thinks that a boy should be named Doug, and the girl's name should start with the letter "M."

> The person who suggested Walt as a boy's name also suggested Deidra as a girl's name; however, this was neither of the two sons.

> The person who suggested Ron as a boy's name did not suggest Dawn as a girl's name.

> The three children are the younger son (who suggested Mark as a boy's name), the one who suggested Dianne as a girl's name, and the one who suggested Sonny as a boy's name.

> Maryanne is one of the girl's names chosen by one of the women.

that human being is no longer potentially a person but *actually* is one. Finally, one might say that achieving the status of being a person is a human being's most mature, self-actualized, fully responsible, and independent state.

Are You the Same "You" Through Time?

In our discussion of personhood above we noted that, according to one view, it is almost as if a human being starts off as a potential person, then develops into an actual person once she has achieved autonomous responsibility as a member of a moral community of persons. Whether you agree with this position or not, the fact of the matter is that a human goes through developmental processes from fertilized egg (zygote), to blastocyst, to embryo, to fetus, to newborn baby, to child, to adult, to lying on one's deathbed, complete with all kinds of steps in between. So, now the question becomes: *What part of you stays the same as you go through the developmental process that is your life?*

It seems as if there must be a permanent part of you, something that stays the same, as time passes by—but what exactly is it? Another way to put it is this: What is it that uniquely identifies you as the same "you" through time? Or, what really comprises your personal identity through time?

Is Your Body "You"?

You could look at pictures of yourself at different points in your life, like when you were five-years old, then ten, then fifteen, then twenty, etc. In those pictures, you note that you are the same person because you recognize that your bodily features are somewhat the same throughout the years. So, you could argue that what stays the same—what identifies you as "you"—throughout time is your body. The same goes for Cousin Johnny, Uncle Albert, or Grandpa Joe, too. You notice features of their bodies—mostly faces—that stay the same throughout the years, and this is what enables you to recognize them as the same persons as time passes. This makes sense to a lot of people.

However, there are problems with this view. One problem has to do with the fact that people get into accidents and live in comas or in a persistent vegetative state (PVS) the rest of their lives, and older people get Alzheimer disease or some other form of dementia. When these kinds of tragic or unfortunate events occur, someone is recognizable as Cousin Johnny, Uncle Albert, or Grandpa Joe in bodily form, but we don't seem to want to say that they are the same persons. Almost no one would have seriously said that the famous Terry Schiavo was the same person in a PVS at the end of her life (2005) as she was when she was twenty-seven, just before she collapsed and lapsed into cardiac arrest in her Florida apartment (1990). So too, Grandma Jean may recognize Grandpa Joe's body as the man she married fifty years ago, but since his Alzheimer has progressed to such a severe level, she'll likely admit that "he's not the same person anymore."

Even if we put aside tragic events like comas and dementias and look at the basic developmental stages of a human's life, we can see that someone's body can change through time to the point of being unrecognizable—yet, when we hear them speak and share thoughts and memories, we immediately recognize them as the same person. Think of the people who go through a weight loss boot camp, like on the show *The Biggest Loser*, for example. We wouldn't say that they are different persons altogether, just from having gone through their wholesale bodily transformations, would we? The same goes for anyone who ages: Their bodies are different—almost wholly and completely different—as they age from five-years old, to ten, to fifteen, to fifty, etc. Again, we would say they're the same person even though bodily they are completely different.

BOX 7-5 CRITIQUE IT!

Someone might try to defend the claim that you are your body by pointing to an individual's unique DNA and cells in the body and argue that these parts of the body are what stays the same and are what identifies an individual through time. However, first, throughout one's life (1) DNA is consistently being altered via regular mutations occurring, and (2) cells are constantly being altered via mitosis and meiosis. So, the body is definitely in flux even at these levels!

And second, identical twins have the same DNA. But clearly they are not the same person. Therefore, as intriguing as this argument might sound on the face of it, it does not work.

Finally, as we have noted above with our science fiction-like examples of artificially intelligent (AI) computational systems and the brain in a jar, it does not seem like a body (except for a brain of some sort) is necessary to be considered a person. Maybe, too, the body is not necessary to identify someone as the same person through time? Although, this may not be correct since, if Frank does have a friend named Fred who is a brain in a jar, someone may argue that it is Fred's *brain* (a bodily part) that identifies him as who/what he is over time. Every time Frank sees Fred, he recognizes Fred as the same brain he has seen time and time again. So, maybe *some kind of material substance on which a mind can depend* is necessary for personal identity over time.

Is Your Mind "You"?

The reality of aging, comas, PVSs, and various forms of dementia—as well as some AI thought experiments—all seem to count against the idea that the body is what identifies you as "you" throughout your life. Many have argued that some aspect of the mind—for example, consciousness—is what makes you be "you" as you move through time. This makes common sense when we think of Terry Schiavo and Grandpa Joe: they are no longer the same persons when their minds are gone (PVS for Schiavo) or are changed in a radical way (Alzheimer for Joe).

So, let's assume, for the sake of argument, that your conscious awareness of yourself is what you can point to as identifying you as *you*, and as you move through time. Then we can say that conscious Terry Schiavo and conscious Grandpa Joe actually become different beings, different things, after they lose their former conscious selves. This makes sense to many philosophers, and doesn't seem to imply any absurdities. (Note, too, that most people would argue that consciousness of any sort still seems to depend upon a properly functioning brain-like-thing in order to exist.)

Conscious Minds in Bodies

Descartes offers a fantastically clever argument to show that what defines you as you is your mind rather than any particular part of your body (and it is also another argument for substance dualism—see Chapter 6):

1. I can conceive of myself existing without a body.
2. Conceivability entails possibility (that is, if I can conceive of some state of affairs, then that state of affairs is really possible—it is a genuine way the world could have been).
3. Therefore, there is some possible state of affairs in which I exist (the real me), but in which I do not have a body.
4. Therefore, a body is not essential to me.
5. Therefore, I am essentially a non-physical being.

If Descartes is right, then what makes you *you* is your mind, not your body (not even your brain in this case).

BOX 7-6 CRITIQUE IT!

Many philosophers are unsatisfied with this argument from Descartes. The primary worry is with premise 2. If it is possible to conceive of something that is not possible, then premise 2 is false. Some examples might be the apparently impossible objects that can be painted, for example, the impossible waterfall of M. C. Escher. If conceivability doesn't clearly entail possibility, the conclusion is not justified by these premises.

Another objection is that, even if Descartes' premises are true and the conclusion follows, it does not mean that Descartes is not contingently stuck to his body in the actual state of affairs. It may be possible for him to be disembodied, but not in a world like ours. Nevertheless, this is an objection to Descartes' substance dualism, and not the idea that what a person is is essentially mental. That conclusion may actually follow from this argument.

BOX 7-7 Philosophy in Film

In the film *Freaky Friday* (2003), mother and daughter swap bodies and they both know who they really are. Interestingly, the famous British Empiricist, John Locke (1632–1704), suggested this kind of switching-bodies thought experiment as a defense of the idea that the conscious self is what counts as a person, rather than a body.

But what happens to *you* if you *lose consciousness*? Imagine you are knocked out for five minutes as a result of being struck by lightning. Are you still the same person, then, before, through, and after the incident? Why think so?

FIGURE 7-1 · Imagine you are struck by lightning.

"Memories... Like the Corners of My Mind"

Because of this lapse-of-consciousness problem, some have responded that what makes you *you* through time has to do with your *collection of memories*, which are comprised of your own experiences. If you think about it, who you are, and what has remained the same about you throughout your life, is your collection of memories: the time you fell off the slide and got a boo-boo when you were four, the time you were stung by a bee at eight, the time you went to the county fair at twelve, your first kiss, your first car, etc.

> ## BOX 7-8 CRITIQUE IT!
> By the way, the position that "who you are is a collection of memories comprised of your own experiences" solves the lightning strike problem. Usually, though not always, if someone survives a lightning strike, that individual can recall all memories prior to the strike. So, even if someone goes unconscious—as when they are sleeping—it does not matter to their self-identity since they can still appeal to their collection of memories as what remains the same "me" through time.

However, what happens if you have an accident (Jeez! There's a bunch of accidents in this chapter!) and can no longer remember who you are, like the character Henry does in the movie *Regarding Henry* (1991). Henry was kind of a jerk who gets shot in the head by a guy robbing a convenience store and, after recovering, he can no longer remember anything about his past jerky life. He starts to learn to live his life all over again and, this time, he's not so much of a jerk. (Getting shot in the head may have been the best thing that ever happened to him, believe it or not!) We could bite the bullet (no pun intended, Henry!), so to speak, and simply say that Henry really is a totally and completely different person because he has lost all of his memories. Again, this is the same kind of intuition we have about Terry Schiavo and Grandpa Joe after their tragic misfortunes; they're simply not the same persons anymore.

Unfortunately, John Locke offers an example that is not so extreme that calls into question the idea that our memories determine our identity. Imagine that when you turn 30, you clearly remember stealing candy at a store when you were 8. Now imagine that when you turn 70, you clearly remember getting a job at 30, but remember nothing about being 8. So, the 30-year-old remembers the 8-year-old, and the 70-year-old remembers the 30-year-old, but now there

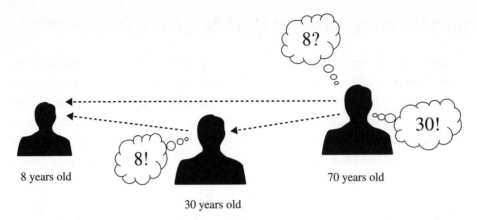

FIGURE 7-2 · Discontinuous memories.

is a discontinuity in your memory: the 70-year-old doesn't remember the 8-year-old. This case is very different from the *Regarding Henry* case. Surely, in this case, the 70-year-old is the same *person* as the 8-year-old, just older and (hopefully) wiser. Locke took this example to be conclusive against the memory account of personhood.

BOX 7-9 CRITIQUE IT!

The most common response to this worry about memory is called the "psychological continuity" theory of identity. The idea is that, even if your memories are discontinuous, your psychological states are not, that is, they all stand in an appropriate causal relationship. The reason the 8-year-old and the 70-year-old are the same person is that the beliefs of the 70-year-old stand in the same causal chain as the beliefs of the 8-year-old.

Locke rejected a version of this view that stated that your consciousness must be uninterrupted over time, to which he responded that sleeping would then result in new people every day. Locke, however, was unaware that our mental states are still causally connected to one another when we sleep, even though we are not directly aware of it.

Interestingly, this probably would not have swayed Locke, since the only things we can know are those directly available to our minds. Therefore, just to *say* that our mental states are held together causally wouldn't hold much weight. No one *perceives* that they are held together this way. This has led to a lively discussion of personal identity over time up to the present.

Are Bodies or Minds Necessary or Sufficient for Personal Identity?

Philosophers like to talk about necessary and sufficient conditions when defining things or explaining events. A necessary condition is a condition, state of affairs, or logical construct whereby, given A and B, to say that A is a *necessary condition* for B is to say that it is impossible to have B without A. Another way to say this is that it is that the absence of A guarantees the absence of B. So, for example, let A be "clouds" and B be "rain." Clouds (A) are a necessary condition for rain (B), or, put another way, the absence of clouds (A) guarantees the absence of rain (B). Notice that there are other necessary conditions for rain, too, like the temperature being a certain degree, atmospheric pressure being just right, etc.

A sufficient condition is a condition, state of affairs, or logical construct whereby, given A and B, to say that A is a *sufficient condition* for B is to say that if A is the case, then B must also be the case. Another way to say this is that is that the presence of A guarantees the presence of B. So, for example, let A be "placing my bare hand in a gallon of water" and B be "my hand getting wet." Placing my bare hand in a gallon of water (A) is a sufficient condition (that's all that suffices) for my hand getting wet (B); or, put another way, the presence of my bare hand in a gallon of water (A) guarantees the presence of my hand getting wet (B). Notice that there may be other events that act as a sufficient condition for my hand getting wet, including placing it in a pool, or in the ocean, or in a pot of cold chicken soup, etc.

Now, let's apply these conditions to our discussion of personal identity over time. Let's start with the idea of a sufficient condition: is a body a sufficient condition for personal identity over time? In other words, is a body all that suffices for personal identity over time? Will the presence of a body guarantee the presence of personhood? Well, the Terry Schiavo and Grandpa Joe examples seem to indicate *no*. Many would say that the bodies in a PVS or ravaged by dementia, laying there in the bed, are merely the emptied shells of what used to house persons. You need more than a body for personal identity over time; you need a mind of some kind.

Is a body even *necessary* for personal identity over time? If it is possible to transfer one's mind into any number of bodies, then it seems that a *particular body* is not necessary for personal identity. Again, think of the *Freaky Friday* possibility. Now, this imaginary, science fiction-y stuff is all well and good; however, even in these thought experiments there seems to be some kind of material substratum on which a mind depends. So, if a *particular* body is not necessary for personal identity (since it's conceivable that the mind can transfer wholesale from one body to another), it still seems that *some kind of body made out of something* is necessary for personal identity insofar as a mind depends upon a material substratum to exist.

What about a mind's being *necessary* for personal identity over time? Well, if what we mean by mind is *consciousness*, then given the fact that people can become unconscious for a time-period, but still retain what seems to be their personhood through some other mental means, consciousness does not seem to be necessary. Recall the fact that someone can get struck by lightning, or get knocked out, or even go into a deep sleep and be unconscious, yet, when they regain consciousness it's their set of memories that seems to give them their personal identity over time.

On the other hand, if mind is understood as a collection of memories comprised of personal experiences or psychological continuity, then it seems that a mind *is* necessary for personal identity over time. What else are you really, but your collection of memories and past experiences or a causal chain of mental events? In fact, this is how the famous British Empiricist David Hume (1711–1776) viewed the human self. In cases where someone has a total and complete memory loss of their former life—as with Henry in the movie, *Regarding Henry*—he argued that they are a wholly new person, building new memories of a new self in a new life.

Now, is mind a sufficient condition for personal identity over time? In other words, is mind *all* that there really is to you as you move through time? If you think it's inconceivable to have a totally disembodied mind, then the answer is *no*. In other words, if you think that some kind of material body—be it a brain in a jar, a borrowed body, a system of silicon and metal (robots), or a plasmatic substance (aliens)—is necessary for the mind to even exist in the first place, then you would have to say that mind alone is not sufficient for personal identity over time. If this is true, you would also have to say that mind *and* body are both at least necessary for personal identity over time.

QUIZ

1. There are those who believe that who you are as a person can change over time, depending on what happens to your body or your mind.
   A. True
   B. False

2. Which of the following is not a criterion for personhood spoken about in this chapter?
   A. Capacity to reason or think logically
   B. Capacity to read minds
   C. Capacity to understand and/or speak a language
   D. Capacity to enter into complex social relationships

3. In fact, if we did not feel justified in drawing an inference about someone's internal cognitive capacities from their external behavior, then the sciences of psychology, sociology, and _____ would cease to exist altogether.
   A. astronomy
   B. astrology
   C. oncology
   D. neurology

4. This is the definition of a person: a being that has all the genetic traits of *Homo sapiens*, as biologists would understand these traits.
   A. True
   B. False

5. Given what we said in this chapter, it doesn't seem that a thing *necessarily* needs to have a brain in order to think, believe, feel, experience, etc., if such cognitive capacities can be *simulated* by other means. So, it seems that a functioning brain, or something that *functions like* a functioning brain, with all of the cognitive capacities associated with such functioning, is completely *insignificant* in determining whether something qualifies as a person.
   A. True
   B. False

6 Given what was communicated about intelligence in this chapter, intelligence is simply not enough for personhood; unless, of course, you think that computers, cats, and chimps are persons. Another way to say this is that the capacity for reason may be _____ for personhood, but certainly not _____ for personhood.
   A. important, significant
   B. significant, important
   C. sufficient, necessary
   D. necessary, sufficient

7. Which of the following is not a social relationship spoken about in this chapter?

    A. Family relationship
    B. Economic relationship
    C. Civil relationship
    D. Virtuous relationship

8. People get into accidents and live in comas or in a persistent vegetative state (PVS) the rest of their lives, and older people get Alzheimer's disease or some other form of dementia. When these kinds of tragic or unfortunate events occur, someone is recognizable as Cousin Johnny, Uncle Albert, or Grandpa Joe in bodily form, but we don't seem to want to say that they are the same _____.

    A. human beings
    B. material beings
    C. both of the above
    D. neither of the above

9. According to the _____ theory, even if your memories are discontinuous, your psychological states are not, that is, they all stand in an appropriate causal relationship.

    A. psychological continuity
    B. psycho-somatic
    C. psycho-somatic continuity
    D. none of the above

10. The absence of A guarantees the absence of B with respect to

    A. a condition.
    B. an absolute condition.
    C. a sufficient condition.
    D. a necessary condition.

chapter **8**

Freedom and Determinism

Have you ever done something that, afterward, you thought, "That was wrong. I shouldn't have done that"? If so, you've probably felt guilty because you know you should have acted differently. But now imagine that you couldn't have acted differently—you didn't act freely. You know it was wrong to do, but you couldn't help doing it in any respect. Should you still feel guilty? Are you morally responsible for doing what you did? In this chapter, we will introduce four types of "freedom" that philosophers discuss and explain some of central arguments about whether we are free and whether freedom is really necessary for being morally responsible for an act.

CHAPTER OBJECTIVES

In this chapter, you'll learn about...

- Different types of freedom
- Alternate possibilities freedom (APF) and moral responsibility
- A deterministic universe
- Determinism and indeterminism
- Compatibilism
- Freedom and moral responsibility
- Harry Frankfurt's contribution to philosophy of freedom

It seems obvious that you're free to raise your arm when you want to, or to make a decision to go on a diet if you want to, or even to choose not to have a baby if you don't want one. But what if you are not *really* free to do these things? What if it just *seems* like you are? What if everything you do has been determined for you already by forces beyond your control, and you are merely under the *illusion* that you are free? This is not some deluded flight of fancy or piece of science fiction. Believe it or not, there are perfectly sane philosophers who offer rational arguments for the claim that no one is free to choose or decide or act—all of your actions are merely the product of natural laws and events that began long before you were born. The bad part is that this possibility raises a significant worry for how we live our lives, particularly for our moral behavior. If we're not free, then how could it be right to hold us responsible for any of our actions?

In this chapter we'll examine several types of "freedom," discuss arguments for the idea that all events in the universe—including human actions—are wholly determined by natural laws, and consider some responses to this conclusion. Finally, we'll consider the possibility that we can still act freely even on the off chance that we are not free to act independently of nature.

Different Types of Freedom

External Freedom

To begin, we need to define and describe what it means for a human being to be free. There are different types of freedom that have been described by thinkers in the history of Western philosophy. There is a kind of freedom whereby a person can do whatever she wishes without any *external* obstacle, blockage, or impediment to the fulfillment of that wish. This can be called *external freedom*. So, for example, if Jane wants a cup of coffee and nothing prevents her from going to Starbucks to get one, then she is externally free since she was able to actually act on, or execute her will for a cup of coffee. On the other hand, if Jim is bound by robbers and thrown in the back of a car, he is not externally free to get up and walk around. So too, people in jail are not externally free to go to Disney World, even if they really want to go, because the penitentiary walls prevent them from leaving the premises. Humans' inability to fly by flapping our wings is another type of obstacle that restricts our external freedom.

Internal Freedom

There is also a kind of internal freedom that is achieved when someone no longer is bound to certain habits, mental attitudes, or emotions. This can be called *internal*, or *psychological*, *freedom*. People can become totally consumed with their own desires in the form of addiction, as when someone can't stop smoking, drinking, eating, or doing drugs. People who are no longer addicted will often claim that they "were slaves" to gambling, cigarettes, fatty foods, or cocaine. People can also become slaves to the effects of external circumstances on their psyche. They have a choice to respond in a "whatever does not kill me, makes me stronger" fashion, or they can cave in to the pressure. If they cave in and become caught in an addiction, this can become a kind of mental slavery; whereas, if they "don't let it get to them," so to speak, they experience internal freedom. Sometimes internal freedom is within our grasp, as when we resist a temptation, but other times it is a result of a psychological disorder, such as obsessive—compulsive disorder or paranoia.

Think of situations where a truly innocent person is wrongly charged with a crime and is placed in prison, and that person sets her mind to remaining strong, accepts the circumstances as they are given to her, and by doing so, maintains sanity and emotional balance. It could be said that, although this person is not externally free, she has achieved internal freedom. Famous figures like Socrates, Jesus Christ, Ghandi, and Martin Luther King Jr.—who all were imprisoned and persecuted at some point in their lives—can be pointed to as having achieved this kind of internal freedom, too.

In the film *Braveheart* (1995), Mel Gibson plays William Wallace, a Scottish knight who wants to gain independence from King Edward I and English rule near the end of the 1200s. The most famous scene is the one where Gibson is on a horse in front of his Scotsmen as they are ready to engage in battle with the English, and Gibson claims, "…they may take our lives, but they'll never take our freedom." This could be looked at as a kind of internal freedom, too. Externally, they were constrained by the British military (denying them political freedom—see below). Internally, they were free, in a sense.

> **BOX 8-1** An Example of Internal Freedom
>
> The famous teacher of Plato and others, Socrates, achieved a kind of internal freedom when he was sitting in jail awaiting his execution by drinking hemlock. Plato recounts Socrates' firm resolve, courage, and sense of freedom in his dialogue *Crito*.

Political Freedom

There is also a kind of freedom which can be called *political freedom*, whereby the citizens of a society have basic rights and privileges, such as "life, liberty, and the pursuit of happiness," they are treated fairly by their governments, and they have equal opportunities to obtain the goods that society has to offer. This is the type of freedom that the British government was trying to deny William Wallace and his kinsmen. This kind of freedom manifests itself when the citizens of a society are free to vote, effectively appeal against an abusive government, and actively engage in the law-making process. Democratic republics, such as the United States, express this kind of freedom in many areas. Recall the following passage from the U.S. Declaration of Independence:

> …That whenever any form of Government becomes destructive of these ends, it is the Right of the People to alter or to abolish it, and to institute new Government, laying its foundation on such principles and organizing its powers in such form, as to them shall seem most likely to effect their Safety and Happiness."

This document is trying to say that in the U.S., people will be politically free to abolish or institute a new government if the old one becomes overbearing. Of course, in reality, the American Confederacy wasn't permitted the freedom described here.

Also, proponents of the French Revolution, as well as most any revolutionaries since, have claimed political freedom as a goal of their revolt. Along with internal freedom, William Wallace and the Scots were striving for political freedom in response to what they viewed as the tyranny of King Edward I.

Alternate Possibilities Freedom

And finally, the freedom with which we are most concerned is called *alternate possibilities freedom* (APF)—sometimes called *metaphysical freedom*. To be free in this sense entails the possibility of choosing one of the many options that lay before you without coercion from nature, genetics, other humans, conditioning, etc. In other words, if you had the opportunity to make a decision all over again, you really and genuninely could have made a different decision. APF is what we are referring to when we say things like, "If I had it to do over, I would have done X instead," and "I really wish hadn't done that." If you didn't believe there was a genuine possibility of choosing your decision on your own, without constraint, it wouldn't make any sense to say you "would have done X instead" or that you regret doing something—without APF, you couldn't have done otherwise. When people talk about *free will*, APF is typically what they're talking about.

So, for example, Jim chooses to take the expressway, rather than Main Street, to get downtown and runs into major traffic. Soon after passing the turn for Main Street, he hears on the radio that there is no traffic on Main Street. He now regrets taking the expressway, and wishes he could go back and make the decision all over again. He really thinks that he *could have taken* Main Street when he considered whether to take the expressway or Main Street. If it's true that he had the alternate possibility of taking Main Street, then it could be said that he was metaphysically free to take Main Street. The fact that we believe we could have done otherwise indicates that we believe we have APF, and not merely external freedom, internal freedom, or political freedom. The chart below highlights the distinctions among the four kinds of freedom (Fig. 8-1).

External Freedom:	Internal Freedom:	Political Freedom:	AP Freedom:
One is free from external obstacles to do something (e.g., coercion, constraint).	One if free from vices, thoughts, beliefs, or emotions that "enslave."	One is free to vote, alter or abolish the government, and actively engage in the law-making process.	One is free to choose among possible actions; if he had the chance to do it over again, he really could have chosen differently.

FIGURE 8-1 · Four kinds of freedom.

Alternate Possibilities Freedom (APF) and Moral Responsibility

Of course, it seems obvious that we are not free to do just whatever we want. Our behaviors are a product of our genetic makeup, our behavioral conditioning, and our individual desires. Could we really *choose* to like broccoli if we don't? Could we really *choose* to be attracted to the opposite sex if we tried really hard? Can you, at 35, work hard enough to compete with a 22-year-old triathlete? None of these seem like genuine options. We're not really free in a lot of respects. So, what's the big deal about being metaphysically free?

Notice that you don't typically hold people morally responsible for their prefence in food, or their animal urges, or their physical limitations. They cannot help these features of their behavior. But what about people who rape or torture? This is where metaphysical freedom becomes important. APF is considered by many philosophers to be the basis for a person's culpability or moral responsibility for actions she has committed or failed to commit. Our belief that we have APF is the reason we praise, blame, forgive, and punish others and ourselves for our actions. It is the reason we regret and train ourselves to do better—because we *really believe* we can do better.

The first three kinds of freedom are important because they can mitigate our responsibility for our actions. If I am not externally free to save a drowning child, APF has been undermined, therefore, I can't be responsible for not saving it. Similarly, if a person is insane (no internal freedom), severely mentally impaired (no internal freedom), or coerced physically (no external freedom) or by political forces (no political freedom) to do something, then we think that person is not morally responsible for his actions; he has been denied the possibility of doing otherwise by some form of constraint. But a Good Samaritan who helps someone change a tire could have chosen *not* to stop, a fireman could choose *not* to respond to an alarm, or a mother could choose *not* to take her infant off the bus with her when she went to the mall. In these cases, APF doesn't seem violated; therefore, we praise or blame the agents accordingly. In addition, every person in a prison somewhere on the planet who has been convicted of a felony (or its equivalent) is in prison because some judicial body (or person) thinks that the person acted incorrectly or immorally *and* that the person *could have acted* correctly or morally to prevent the felony from occurring in the first place. In other words, the judicial system assumes that, if APF has not been violated, a person is legally responsible for her actions.

Therefore, a proponent of APF might offer an argument like the following:

1. In order to hold a person morally responsible for an action, that person must have had APF in some set of circumstances relevant for that action.

   (We cannot say "must have had APF in the moment of the action," because a drunk driver who kills someone is externally prohibited from not killing that person. But the driver is the one who introduced the obstacle by getting in a car drunk. Therefore, the driver had APF in the relevant circumstances.)

2. <u>Some people should be held morally responsible for their actions.</u>

3. Thus, some people have APF.

If someone does not have APF in a set of circumstances relevant for the action, then we simply cannot hold that person morally responsible. This explains our

For example, if you go to the gym regularly (decision, intention, behavior), this is partly because of your desire to be healthy (motivation) which itself is determined by your brain chemistry, partly because of physiology (nature stuff), and partly because of the fact that you have seen others workout, or had a gym class, or had someone tell you the benefits of working out (nurture stuff).

Now, the important philosophical questions are

1. **Is there any way of acting *outside of* or *against* these deterministic processes?**
2. **And, if not, am I still responsible for my actions?**

Hard Determinism

According to the view known as *Hard Determinism*, all events have been determined to occur by previous events, and this includes every human thought, decision-making process, motivation, and action. Be careful, here: Determinism is not Fatalism. *Fatalism* is the view that some event will happen in the future *no matter what happens* between now and then. Determinism is the view that every event has a complete and total explanation in terms of previous events and laws of nature. So, the determinist agrees with the fatalist that the future is already set, but disagrees that it can happen just any old way. For the determinist, the future will happen just one way—and that way is determined by a combination of the past and the laws of nature.

According to hard determinists, we think we are free to act, but it's really just an illusion. One consequence of this view is that: If you are not in control of your actions, it is difficult to see how you could be responsible for your actions. It sounds absurd to imagine a criminal standing before a judge arguing: "Your Honor, I have been determined by the past and laws of nature to be the conduit through which this action occurred. I did not choose it and I could not have chosen otherwise. Therefore, it would be unreasonable for you to blame or punish me for this crime."

One response to this picture is to agree with hard determinism but claim that something like Behaviorism (see Chapter 6) is the appropriate way to understand our praising and blamimg behavior. If we want a person to stop being a criminal, we do not morally blame them, but we may want to change her *motivations* or *neuroanatomy* (see Fig. 8-6). We can't change her DNA, nor can we go back in time to alter the historical aspects of her nurturing, but perhaps we can:

punish and counsel her so as to change her motivations; for example, by placing her in prison with a chance of reform from vicious motives (hate, envy, sloth, overindulgence, etc.) to virtuous motives (love, righteousness, honesty, hard work, prudence, etc.). Or, as with our above gym example, get her to see that being fat and unhealthy is something she doesn't want, thus changing her motivation to act.

give her neuro-pharmacological drugs that affect her mood which, in turn, will affect her behavior. For example, we might give an overly anxious criminal a neurotransmitter-regulating drug that calms him down so that he no longer has the desire to rape. Or, again with our gym example, give someone a diet pill, caffeine pill, or some other neurotransmitter-regulating drug that peps him up.

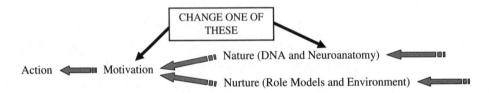

FIGURE 8-6 • Two ways to change behavior in a deterministic world.

The problem is that even these solutions presuppose that someone can choose freely to act in such a way to change another person's motivations. But if everyone's actions are determined, then I either will act to change someone's motivations or I won't. I will either evaluate someone's behavior as a Behaviorist or I won't. And because that person's actions are determined, too, their motivations will either be changed or they won't. There will not be any "deciding" that means anything at all apart from the neurological and behavioral processes that produce decisions.

Hard Indeterminism

Some philosophers are not satisfied with hard determinism. They argue that we have significant evidence that we can act independently of our cultural and biological conditioning. These philosophers hold a view known as *Hard Indeterminism*, which is the view that at least one kind of event in the universe is *undetermined* in a way sufficient for moral responsibility, namely, human action. This indeterminism exists principally because people have APF. The idea is that there is a unique kind of event within causal history distinct from purely deterministic and purely indeterministic events; that is, events caused by "first movers," agents who can act independently of natural laws and causal chains. So, whereas every other kind of event in the universe has a history that some- one with an all-knowing "god's-eye-view" could trace and determine, *human* events (human acts) are different in that they are not determined and the human committing the act is the beginning agent of that action.

If we look again at a diagram showing the motivations of human action (Fig. 8-7), the indeterminist would maintain that freedom of choice exists somewhere in between the motivation and action, such that there is the pos- sibility of choosing any number of alternative actions. Another way to think of this is that the determinist's chain is broken (or broken into) by free human choice. And, again, this commonsense view is what most people take to be the freedom necessary for moral responsibility. The person who is motivated to go to the gym could act *against* that motivation, for example, and instead of exer- cising his pectorals and biceps at the gym, he could lay on the couch exercising his jaw and throat muscles by eating potato chips and ice cream.

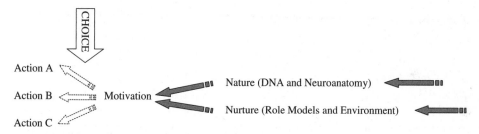

FIGURE 8-7 · The influences on your decisions in an indeterministic world.

In fact, that people constantly maintain that they act *against* their motiva- tions would be further support for the indeterminist since the deterministic causal chain between a certain motivation and a certain action has been broken.

Of course, the determinist could always respond that the person still has been determined to act because all that has happened is that one motivation has been replaced by another so that, instead of being motivated by *A* deter- ministically to do *X*, someone is now motivated by *B* deterministically to do *Y*! So, there's no real free choice at play when someone goes to the gym, or when he plays couch potato; it's just different motivations doing their deterministic thing.

> ### BOX 8-3 CRITIQUE IT!
>
> It has been said that the hard indeterminist has the new problem of directly linking a free will to a person acting because there seems to be no strong connection between the two. How does some free will that comes out of nowhere, so to speak, affect material bodies in a material universe? Also, this free will—which no one can see—supposedly having an effect on a human body all seems so mysterious and random, too! And, if there is no strong connection between the person and the action, then there is no way to hold that person responsible for an action committed, and, hence, no grounds for praise or blame.

Compatibilism

A *compatibilist* tries to reconcile—or make compatible—the determinist's view of the universe with the freedom required for moral responsibility. *Compatibilism* is the view that a person can be morally responsible for her actions even if determinism is true. Compatibilists are not hard determinists; they do not claim that, in fact, nature is determined. They say merely that, *if nature were* deterministic, it is still possible for people to be morally responsible for their actions. Compatibilists argue that APF is *not* the type of freedom necessary for moral responsibility. In fact, you can be morally responsible for your actions even if you could not have done otherwise. But if APF is not the sort of freedom required for moral responsibility, what could be?

Some compatibilists defend a view about a person's attitude toward the actions they commit. For instance, if a person "owns" the action psychologically, that is, assents to it, feels no remorse over it, is motivated by the right sort of factors, then that person is responsible for the action even if she couldn't do otherwise. This condition of "owning" an action refers to what are called the agent's "reactive attitudes." The appropriate reactive attitudes toward an action determine moral responsibility independently of whether the agent could have done otherwise.

This view would accommodate most criminal acts—the willing commission of a crime, the intentional telling of a lie, the desire to punch someone who is antagonizing your significant other. In the rare cases where someone commits a crime and feels just absolutely horrified at having done so, and wished they could have done otherwise, it is likely that his action was not motivated by the right sort of factors—perhaps mental illness, drug abuse, or physical abuse spawned the wrong sort of response to normal stimuli. In these cases (even according to many who accept APF), it would seem that the person deserves treatment rather than punishment.

But still, something seems amiss. If determinism is true, even a person's reactive attitudes are out of her control. She can't even "own" her own feelings, much less her actions—both are hers because of the unlucky fact that she was in one causal chain of events rather than another. Its wildly counterintuitive nature has led to many criticisms of compatibilism. But one defense of the view stands out as particularly difficult to avoid.

Harry Frankfurt's Counterexample to the Principle of Alternative Possibilities

We have noted that APF entails the possibility of always being able to act other than one has acted if one had the opportunity to perform the action over again. As we have seen, many philosophers have associated APF with moral responsibility. This has been formulated into a principle called the *principle of alternative possibilities* (PAP), which states that: "A person is morally responsible only if he could have done otherwise." But a famous philosopher named Harry Frankfurt (b. 1929) challenged the legitimacy of (PAP) in such a way that many philosophers have given it up altogether and have become compatibilists. In fact, because of this argument, the majority of philosophers working in free will are now compatibilists.

The argument was presented in the form of a counterexample that has come to be known as the "Frankfurt Case," and the variations, "Frankfurt-style Counterexamples." Here is one popular version:

> Imagine that a malevolent neuroscientist, Ben, wants to kill Sam, but doesn't want to get caught. Luckily, Ben discovers that Jim also wants to kill Sam. Ben decides to place a chip in Jim's brain so that, if Jim decides not to go through with the killing, Ben can press a button and cause Jim to intend to kill Sam. But if Jim goes on with the killing, Ben won't have to do anything. Ben places the chip in Jim's brain and waits. As it turns out, Jim decides *on his own* to kill Sam and carries out the plot. In this case, Jim kills Sam, Jim could not have done otherwise (because then Ben would have pressed the button and caused Jim to kill Sam), and Jim seems morally responsible for Sam's death.

If this is right, Jim is morally responsible *and* he could *not* have done otherwise! If this is right, then (PAP) is false, and APF is not the relevant sort of freedom for moral responsibility. Pretty powerful stuff.

The Intentions Response

Believe it or not, there have been whole books written around this kind of counterexample. However, one response goes like this: In order for Ben to know that Jim would go through with the murder, Ben had to know something about Jim's motivations/intentions. So, if Jim had changed his mind, and intended not to kill Sam, then Ben would have intervened. If Ben had intervened, everyone agrees that Jim would not have been morally responsible for Sam's death (though his body carried out the deed, he was *mentally* coerced because of the chip in his brain). But now, consider this: even though Jim could not have done other than kill Sam, he could have done other than he did, namely, he could have *intended* to do otherwise. This change of intention would have removed any of Jim's culpability/responsibility for killing Sam, since then he would have been coerced to do it. Therefore, Jim is morally responsible for killing Sam because he killed him *on his own*. He could have done otherwise, namely, he could have brought it about that Ben killed Jim using his body.

Most philosophers find this response convincing against the traditional Frankfurt Case. But stranger and more complicated examples suggest that this response may not be good enough to cover all Frankfurt-style Counterexamples.

Still Struggling

The point of the Frankfurt-style Counterexamples is to challenge our common sense belief that it is the fact that we could have done something other than we did that makes us morally responsible for our actions. Recall an example from first Still Struggling? box in this chapter: at noon tomorrow, nature has determined that you will slap a baby until it cries and you will condone the act. In this case, your decision-making procedures have been coerced, nonetheless, argue compatibilists, you are still responsible for slapping the baby. What is it that makes you responsible? Compatibilists disagree about the precise necessary conditions, but many argue that it is a matter of your *attitudes* toward the act or reasons for acting—if you want or approve of slapping the baby or think you have good reasons to do so, then you are morally responsible for the act; if you don't, you're not. Some compatibilists link moral responsibility to a *pattern* of such attitudes, especially as they relate to reasons that you perceive as motivating you to act.

One problem with these compatibilist approaches is that they have trouble explaining the difference between two types of agents: (non-disabled) agents who genuinely exhibit a pattern of immoral behavior and don't regret it (e.g., cartel leaders, mafiosos, etc.), and mentally disabled agents who exhibit a pattern of immoral behavior and don't regret it (e.g., schizophrenics). Since both types of agents exhibit a pattern of behavior consistent with the motives they perceive for acting and condone the resulting actions, it would seem that "attitude" compatibilists face a dilemma: they must conclude either that (a) members of neither group are morally responsible because none are appropriately responsive to moral reasons, and therefore are not moral agents, or that (b) members of both groups are morally responsible for their actions because all respond consistently and approvingly to the reasons they perceive. But clearly the first group is morally responsible and the second isn't.

Are You Even Free to Read On?

At this point in your reading, we (Rob and Jamie) would contend that you are wholly determined to continue reading on because we are such good writers and you can't help but want to continue reading on! We're just kidding, of course. But we hope that we have at least gotten you to question the common-sense view that you are wholly free to do what you want when you want to do it, as well as how it is that freedom and moral responsibility seem to go hand in hand with one another. Go ahead—you're free to read on…

QUIZ

1. If we're not free, then how can we be held responsible for _____?
   A. what other people do
   B. all of what other people do
   C. any of our actions
   D. None of the above

2. This kind of freedom entails that a person can do whatever they wish to do, without some kind of obstacle, blockage, or impediment to the fulfillment of what that person wants.
   A. External freedom
   B. Internal freedom
   C. Political freedom
   D. Alternate possibilities freedom

3. The fact that we think we could have done otherwise is an indication that we believe in the reality of
   A. external freedom
   B. internal freedom
   C. political freedom
   D. alternate possibilities freedom

4. Speaking of regretting decisions we have made, _____ is considered by many philosophers to be the basis for a person's culpability or moral responsibility for actions that she has committed or failed to commit. Our belief that we have this kind of freedom is the reason we praise, blame, forgive, and punish others and ourselves for our actions.
   A. external freedom
   B. internal freedom
   C. political freedom
   D. alternate possibilities freedom

5. According to a determinist, your motivations are determined only by your environmental influences—nurture stuff like role models, other opinions and ideas, your general upbringing, etc.
   A. True
   B. False

6. According to _____, all events have been determined to occur by previous events, and this includes every human thought, decision-making process, motivation, and action.
   A. Determinism
   B. Fatalism
   C. Determination
   D. None of the above

7. According to hard determinists, we think we are free to act, but it's really just _____.
   A. an act
   B. a false action
   C. an illusion
   D. an illusion that's not really an illusion

8. Hard Indeterminism is the view that at least one kind of event in the universe is wholly *undetermined*, namely, _____.
   A. God's actions
   B. animal instincts
   C. human anatomical functions
   D. human action

9. _____ argue that APF is *not* the freedom that is necessary for moral responsibility. In fact, you can be morally responsible for your actions even if you could not have done otherwise.
   A. Compatibilists
   B. Incompatibilists
   C. Determinists
   D. Hard determinists

10. Most philosophers find the intentions response to be convincing against the traditional Frankfurt Case.
    A. True
    B. False

The Question of God's Existence

What do people mean when they use the term, "God"? Can anyone "prove" that a being like god exists? What would it even mean to offer evidence that something like god exists? In this chapter, we'll introduce the philosophical discussion surrounding the existence of a god, in particular, a very powerful being who created the universe, often denoted by capitalizing the first letter: God. We'll also introduce the primary philosophical approaches to arguments for and against the existence of God, and then we'll explain three traditional arguments for God's existence and one against.

CHAPTER OBJECTIVES

In this chapter, you'll learn about...

- The concept of "God"
- Philosophical approaches to god's existence
- Arguments for god's existence
- An argument against god's existence: the problem of evil
- The logical and evidential problems of suffering

The Concept of "God"

Philosophy of religion is a branch of philosophy that seeks to apply the tools of philosophy (logic and language) to questions about religious claims. Religious claims are claims traditionally associated with one or more particular religious traditions. These might include claims about various gods or goddesses or other spiritual realities; various revelations of one or more of these realities (i.e., their sacred texts); claims about humans and their relationships to one or more of these realities, including moral codes, the possibility of life after death, the conditions required for having a life after death, etc. Philosophers take these claims and ask philosophical questions about them: Why believe gods or a God exists? What sort of being must a god be like (can it make a rock so big it can't lift?)? How reliable is a particular sacred text? How likely is it that humans can live without their physical bodies? etc.

In this chapter, we will be concerned with one specific question in philosophy of religion: *What reasons are there for believing that a certain type of god ("God," if you will) does or does not exist?* What sort of "god"? Traditional Western arguments for god's existence (whether from the Jewish, Christian, or Muslim tradition) presume that the kind of divine being we would care about is a being that is all-knowing (omniscient), all-powerful (omnipotent), and all-good (omni-benevolent)—that is, "God" with a capital "G." However, philosophers from these traditions are careful to note that their burden is to prove that such a being exists, not to presume it exists before any evidence has been offered. They therefore begin by investigating whether any type of being exists that might be an even remotely plausible candidate for God, for instance, a creator deity, a designing deity, or a deity that could not possibly *not exist*. If no one can offer arguments that support the existence of this sort of being, then the plausibility of a much more significant being, God, is very low. On the other hand, if a creator, or designer, or necessary being exists, then the traditional conception of God is much more plausible, though, by no means certain. If philosophers of religion establish that some divine being exists, they acknowledge they must still go further and argue that their particular conception of God exists (e.g., the God of Judaism, the God of Christianity, or the God of Islam, etc.).

Some philosophers of religion are interested, not in supporting the claim that God exists, but in disproving the existence of such a being. They offer arguments either showing that traditional arguments for God's existence do not work, or that the idea of God is implausible or even impossible. The most common argument against the existence of God is that there is too much evil or suffering in the world to think that such a being is likely to exist. We will explain some common approaches to the question of God's existence, consider three arguments for God's existence, and, finally, we will discuss the "problem of evil" argument against God's existence.

Philosophical Approaches to God's Existence

Approaches to the question of whether God exists can be divided among two categories: evidentialism and non-evidentialism. *Evidentialism* is the view that belief or disbelief in God requires objective evidence. *Objective evidence* is evidence that everyone involved in the discussion could have access to. It is evidence I can share with you. Even if only a few physicists in the world have seen an image of an electron in an electron microscope, it is still the sort of thing every human could see, and another physicist could double-check the results. Objective evidence stands in contrast to *subjective evidence*, which is evidence that only single individuals have access to. For instance, if you are in pain, you can tell me *about* it, or I can watch your behavior and recognize that *you behave as I behave* when I'm in pain; but I cannot experience *your* pain. I do not know what *your pain* feels like, and you don't know what *my pain* feels like. Therefore, pain is a subjective experience. In the same way, religious experiences are subjective. I could have one, be fairly confident that I've had one, and yet not be able to share that experience with you. Thus, an evidentialist would not accept religious experience as sufficient evidence of God's existence.

There are three prominent versions of evidentialism: theism, atheism, and agnosticism. *Theism* is the view that evidentialism is true and there is objective evidence for God's existence. Theists offer arguments that they believe provide objective evidence sufficient for believing that God exists. We will discuss three theistic arguments in this chapter. The term "theism" is also used more broadly to include anyone who believes in God. For our purposes, we will only use it to refer to evidentialists who believe there is objective evidence for believing in God.

Atheism is the view that evidentialism is true and that there is objective evidence against God's existence. Atheists offer arguments that they believe provide objective evidence sufficient for believing that God doesn't exist. Some argue that atheism more properly applies to people who do not accept the arguments for God's existence or have no belief whatsoever with respect to the question of God's existence. This definition, however, becomes absurd when we realize that it would force us to call infants and the mentally handicapped atheists. The categories of theism, atheism, and agnosticism, refer to types of belief. And people who have formed no opinion at all on the question of God's existence, because of physical immaturity, physical handicap, or simple lack of exposure cannot be categorized as holding any particular belief about evidence for God's existence. In addition, given this definition of atheism, we could not properly distinguish atheists from agnostics, a third version of evidentialism. We will discuss one atheistic argument in this chapter.

Agnosticism is the view that evidentialism is true and that there is no objective evidence sufficient for believing either that God exists or that God doesn't exist. Agnostics have weighed the evidence and believe either that the evidence for both views is insufficient in some important respect, or that the evidence for both sides is equally strong, so that the only rational position is to suspend judgment until more evidence for one side or the other is offered.

Some philosophers argue that evidentialism is too strong a criterion for belief in God. Evidentialism may be appropriate for subjects such as history or physics, but not for matters of religion or morality. These philosophers accept a version of *non-evidentialism*, which is the view that belief or disbelief in God does not require objective evidence. There are three prominent versions of non-evidentialism: fideism, prudentialism, and subjectivism. Be careful not to think of these as counterparts to the three versions of evidentialism (where the first is held by people who believe in God, the second is held by people who are unsure, and the third is held by people who do not). Versions of non-evidentialism are different than the versions of evidentialism in that someone who holds any one of them may believe that God exists, be unsure what to believe, or believe that God doesn't exist. For instance, *fideism* is the view that non-evidentialism is true and that belief in God depends on whether a person has faith that God exists, not on any evidence that person may have that God exists. A fideist may have faith or not. If the fideist does not have faith, that person does not believe in God. On this view, it is perfectly reasonable for some people to believe and for others not to believe—this is because some people have faith and others do not.

Similarly, *prudentialism* is the view that non-evidentialism is true and that belief in God is best justified on prudential grounds, that is, on whether it is in my best interests to believe in God. Prudential reasons are not reasons to believe that the claim, "God exists," is true, but merely reasons to believe the claim, "God exists," even if you have no idea whether it is true. For instance, twentieth-century American philosopher William James argues that, for some people, the idea of God is necessary for their emotional well-being, or to explain the unexplainable, or for believing in some sort of foundation of morality. James argues that these people are perfectly justified in believing that God exists, though these reasons are not evidence that the claim, "God exists," is true. James himself is also prudentialist, though he claims that, since he has no need for the concept of God, it is perfectly consistent for him to believe there is no God on prudential grounds, regardless of whether God actually exists.

And finally, *subjectivism* is the view that belief in God can be justified on purely subjective evidence. It may also be justifiable on objective evidence, but subjective evidence, such as a religious experience of God's presence, is sufficient for justifying the belief that God exists. A person who does not believe that God exists may also claim to hold a version of subjectivism. He may say things like, "I would have to see for myself," or "God would have to reveal himself to me before I could believe." Since he would regard subjective experience as sufficient for believing, though he hasn't had any subjective experience (and he may be skeptical of others who claim to have had religious experiences), he is a subjectivist who doesn't believe God exists.

Some philosophers argue that only subjective evidence is sufficient for belief that God exists. For example, the theologian John Calvin argued that people can believe in God only if they have experienced the "internal instigation of the Holy Spirit," which is a subjective experience. Figure 9-1 summarizes all of these approaches to the question of God's existence and will hopefully be an easy way for you to refer back to these views as you wrestle with the evidentialist arguments for and against the existence of God.

BOX 9-1 Critique it!

Where would infants and the mentally handicapped fall within these categories? Or imagine someone raised in a purely secular home, where the question of the existence of a higher power has never even been raised. Apparently these individuals have no beliefs regarding God whatsoever, and since these categories identify various beliefs about the evidence required for belief in God, it is possible that these categories do not represent everyone.

However, philosopher of religion William Rowe draws a distinction among three degrees of "agnostic": weak, moderate, and strong. The "weak agnostic," though she understands, or could understand, the concept of God, simply has no beliefs with regard to God whatsoever. The "moderate agnostic" is someone who believes that the question of God's existence is not one humans have the capacity to answer, so he suspends judgment. And the "strong agnostic," is someone who is unconvinced by any arguments for or against God's existence and claims that evidence is relevant to the question—that is, only someone with appropriate evidence should believe (William Rowe, "Agnosticism," *Routledge Encyclopedia of Philosophy,* London: Routledge, 2008).

Therefore, the sense of agnostic we have identified is Rowe's strong agnostic. In addition, infants, the mentally handicapped, and the pure secularist are most likely weakly agnostic in Rowe's sense.

Evidentialism: the view that belief or disbelief in God requires objective evidence

    Theism: the view that evidentialism is true and there is objective evidence for God's existence

        Some theists:
                Aristotle; Aurelius Augustine; Thomas Aquinas; Rene Descartes; John Locke; Alvin Plantinga

    Atheism: the view that evidentialism is true and there is objective evidence against God's existence

        Some atheists:
                Epicurus; Lucretius; Friedrich Nietzsche; Karl Marx; Richard Dawkins

    Agnosticism: the view that evidentialism is true and there is no objective evidence sufficient for believing that God exists or doesn't exist

        Some agnostics:
                Michel de Montaigne; Machiavelli

Non-Evidentialism: the claim that belief or disbelief in God does not require objective evidence

    Fideism: the view that non-evidentialism is true and that belief in God can only be justified in virtue of having faith in God

        Some fideists:
                Soren Kierkegaard (tradition often places him here); Blaise Pascal (tradition often places him here)

    Prudentialism: the view that non-evidentialism is true and that belief in God is best justified on prudential grounds

        Some prudentialists:
                Blaise Pascal (more plausible); Immanuel Kant; C. S. Peirce; William James; John Dewey

    Subjectivism: the view that non-evidentialism is true and that belief in God can be justified by subjective evidence (often personal religious experience)

        Some subjectivists:
                Soren Kierkegaard (more plausible); John Calvin; John Wesley

FIGURE 9-1 • Table of approaches to God's existence.

Still Struggling

You may be wondering why anyone would care to distinguish these approaches to the question of God's existence. It is crucial to understand why they are important. Where you fall on this chart will determine how you interpret various arguments for and against God's existence. For example, if you are a non-evidentialist, then evidentialist arguments will be largely irrelevant to your beliefs about God. You may find the arguments to be fun logic puzzles, but the conclusions will have little effect on your actual beliefs. On the other hand, if you are an evidentialist—that is, if you believe that, for all of your beliefs, you should follow the evidence where it leads—then the arguments for and against God's existence will be very important for you. You will examine the evidence and follow it where it leads, whether to theism, agnosticism, or atheism.

Arguments for God's Existence

Two important questions we should ask before considering the arguments for or against God's existence are (1) what does the argument attempt to establish? and (2) how strong does an argument have to be for me to accept it? Many students of philosophy approach the question of God's existence presupposing an answer to question (2), and do not ask (1) at all. In addition, the answer they presuppose for (2) is: *conclusive proof.* Conclusive proof may be a useful criterion for many areas of study, e.g., theoretical mathematics, pure logic, Sudoku. But it may be too strict for many other areas, e.g., how species evolve, whether the medicine a doctor gave me will help or harm me, whether this plane I'm on will crash.

Consider the example of getting on a plane. When you travel by air, you put your life in the hands of two or three strangers who are driving a large metal building several hundred miles an hour at around 30,000 feet above the ground. You will perhaps agree that whether a plane will crash depends on many relevant factors including, how much sleep the flight and mechanical crew have had, whether the flight and mechanical crew are drunk or high or tweaked, the mechanical history of the plane, whether there is a terrorist on board, the weather that day, and whether there are any large flocks of birds on a course to intercept the plane. But we would venture to guess that you checked very few, if any, of these factors before you last boarded a plane. Your belief that the plane will not crash is based almost wholly on one statistic: *planes arrive safely more often than not*. This is hardly conclusive proof. In fact, this statistic has *nothing* to do with the flight you boarded. That statistic refers only to past flights. What about *this* plane?

This example is not intended to convince you to stop flying, but merely to point out that your standard of evidential proof for activities that involve whether you live or die is fairly low. Perhaps you had never thought about this and now feel that your standard of evidence should be raised. Or perhaps you are perfectly content with your standard of proof. Regardless of where you find yourself, this makes a difference for how you evaluate the strength of arguments for and against God's existence. If your standard is too high, it is unlikely that any argument could convince you one way or another. If your standard is too low, you may be too easily convinced, or find the arguments on both sides equally convincing even when one is clearly stronger than the other. Questions about the appropriate degree of evidence for philosophy of religion are too involved for our discussion here, but keep in mind that this is an important issue in philosophy of religion.

Even if you haven't decided how much evidence it would take to convince you, you can still evaluate whether an arguer accomplishes his task by knowing the answer to question (1), what he intends the argument to accomplish. Someone who offers a deductive proof (a sound argument) for or against God's existence faces a much more difficult task than someone who offers an inductively forceful argument (a cogent argument). Similarly, someone who only attempts to defend the existence of a creator deity faces a much easier task than someone who attempts to justify belief in, say, the Christian God. And someone who presents an argument for a creator but claims that she has proven the existence of the Christian God is clearly claiming too much—there is much more to the Christian conception of God than merely "creator." So make sure you understand exactly what each arguer attempts to accomplish, and then evaluate whether he achieves this goal.

The Teleological Argument

The teleological argument is an evidentialist argument for God's existence that appeals to indicators of apparent design or purposefulness in either the universe in general or among biological organisms in particular as evidence for the existence of a divine designer. The word "teleological" comes from two Greek words *telos*, which means "purpose" or "goal," and *logos*, which means "study of" or "logic of." So, "teleological" means the study of something's purpose. The idea is that some features of nature are so complex and so well-suited to the events they bring about that it is difficult to believe they could have been caused by a string of random events; therefore, a divine being, often called an "intelligent designer," must be responsible, and, therefore, exist.

A basic version of this argument looks like this:

1. Some features of nature are incredibly complex and apparently organized to perform a very specific function (e.g., the mammalian heart and eye, the replication of genes, the precision of natural laws).

2. These features are best explained either by pure chance or by the work of some incredibly intelligent, incredibly powerful being.

3. It is unreasonable to believe this sort of complexity and precision is a result of mere chance (just as it would be unreasonable to believe that an iPhone or a PC, or even a steel hammer, was produced by a series of random events).

4. Therefore, the best explanation for the complexity and precision of these features of nature is that an incredibly intelligent, incredibly powerful being exists to organize them this way.

Notice, first, that this argument is inductive. It does not purport to offer conclusive proof that an intelligent designer exists. It concludes only that this is the *best explanation* for some features we find in nature. Second, notice what the argument does not establish: It does not conclude that this being is all-knowing, only that it is incredibly intelligent; it does not conclude that this being is all-powerful, only that it is incredibly powerful; and it does not say anything about whether such a being is morally good. So, even if this argument is successful, it is a long way from establishing the traditional God of Judaism, Christianity, or Islam. This doesn't mean that Jews, Christians, or Muslims cannot use this argument, only that they would also need additional arguments to establish the further conclusion that *this* intelligent designer is the God they worship.

The most famous version of this argument was offered by Anglican Bishop William Paley in 1802, and is known as the "Watchmaker Argument." Paley asks you, the reader, to imagine walking along a heath (a section of elevated ground characterized by a wide variety of low shrubbery) and coming across a watch on the ground. He says any answer you give to a question about the origin of this watch should be distinctly different from the answer you would give to a question about the origin of any stone you might find on the heath. The stone, for all you know, might have lain there from eternity, or have been carved by wind and sand off of a larger rock. The features of the watch, however, seem to suggest a much richer past, *even if the watch is broken*. Something about the materials of the watch, its precisely interrelated parts, and the organization of these parts leads you to believe immediately that some intelligent being was responsible for the watch's design and its placement on the heath.

This is not a far-fetched line of reasoning. In fact, it is the way archaeologists determine whether a stone found at an archaeological dig is a tool made by human hands for a purpose or just a rock. The features of the stone help scientists decide; and a watch is a more complicated and obvious example than many of the stones that archaeologists claim to be tools. With this strategy in mind, Paley then asks you to consider the mammalian eye, with its unique materials, precisely interrelated parts, and organization for the purpose of seeing. Now simply ask: Is the eye more like the watch or the stone? Paley concludes that it is more like the watch, and since you conclude that a watch must have an intelligent designer, so you should conclude the same for the eye.

Is this a good argument for the existence of something like God? If it were true that our only choices for explaining complex natural organs like the eye were chance and a being like God, it seems plausible that God is a better explanation. However, just a few years after Paley offered his Watchmaker Argument, Charles Darwin published the first edition of, *On The Origin of Species* (1859). In that work, Darwin offered a third possible explanation for biological complexity: evolution by natural selection. To be sure, the idea of biological evolution had been around long before Darwin, but most accounts of it depended so heavily on chance or on believing that every creature had a predetermined essence (a theory of Aristotle's), that no particular version was widely accepted. The virtue of Darwin's account is that it involves a non-chance, natural mechanism by which very complex organisms can develop as they interact with a variety of environmental pressures (predators, food sources, climate, etc.). Natural selection involves chance (in the mutation of genes and changes in environmental pressures), but it is not purely chance, since the genetic code of an organism constrains the types of changes over time that are possible.

To be sure, Darwin had no idea about genetics. Mendel was developing his famous theory of genetics with his pea plants around the same time that Darwin was working on his theory of natural selection. Unfortunately, the two never met and it would take decades before natural selection met genetic variation. Nevertheless, Darwin's idea of cumulative selection on the basis of gradual changes over time set the stage for a revolution in biology that was corroborated by discoveries in physics (laws of energy conservation), geology (plate tectonics and radioactive dating), genetics (Mendel's handiwork), and paleontology (the discovery of various transitional forms, such as the archaeopteryx).

The trouble that Darwin's theory wrought for the above argument is that it changes premise 2. Instead of reading:

1. These features are best explained either by pure chance or by the work of some incredibly intelligent, incredibly powerful being.

It now has to read:

2. These features are best explained by pure chance, by the work of some incredibly intelligent, incredibly powerful being, or by natural selection.

Premise 3 remains true: It is unreasonable to believe the eye developed on the basis of mere chance. But Darwin's theory does not depend on pure chance. Chance factors operate on an organism at a time (in the form of genetic mutations and changing environmental pressures). But the features of the organism produced by genetic changes provide a sort of limit and even a direction for the chance factors. Darwinian evolution is not a pure chance process; it proposes that species change cumulatively over long periods of time, which results in a wide variety of biological organisms. Thus, many philosophers conclude that natural selection is also a better explanation of biological phenomena than chance.

The question, now, is whether it is also better than an intelligent designer. Most philosophers think so. Natural selection is a simpler explanation (it appeals only to forces we already knew were at work, namely, genetics and the environment; and it doesn't require an extra supernatural force), and it has more explanatory power (it can fill in some details of "how" evolution occurred, whereas we can't ask God how he did it). On the basis of theoretical virtues like these, it would seem that natural selection is a better explanation than design.

BOX 9-2 Even if You Don't Buy Evolution

Notice that this argument goes through even if you don't think Darwinian evolution is a very plausible theory of biological diversity. Even if there were no evidence supporting the theory. A theory that offers a non-chance, non-design explanation sufficient for explaining biological diversity can be evaluated against competing theories on its theoretical virtues alone.

Theoretical virtues are descriptions of how an explanation is related to some phenomenon. For instance, the number of mechanisms or processes that an explanation requires to explain a phenomenon is its *simplicity*. And the number of phenomena it can explain is its *explanatory scope*. While the design hypothesis has a very wide scope—it can explain just about anything—it has very shallow *explanatory depth*, that is, the details of how the phenomena is produced on the explanation. Natural selection, though having a narrower scope, is explanatorily deeper than design, because it explains how nature produces diversity. In addition, natural selection is simpler in that it appeals only to the features of the world everyone already accepts (organisms, genes, environmental pressures), where as the design hypothesis includes all of these plus a supernatural agent. Therefore, even if you are not convinced that evolution is very plausible, it is still possible to conclude that it is a better explanation for the apparent design we perceive in nature than a supernatural being.

This is by no means a conclusive argument against an intelligent designer. But it does suggest that we do not need to appeal to such a being to explain complex biological phenomena. If this is right, then Paley's argument will not work for complex and apparently purposeful biological phenomena. The argument would now end like this:

3. It is unreasonable to believe this sort of complexity and precision is a result of mere chance (just as it would be unreasonable to believe that an iPhone or a PC, or even a steel hammer, was produced by a series of random events).

4. It is more likely that biological complexity is a result of natural selection than an intelligent designer.

5. Therefore, the best explanation for the complexity and precision of some features of nature is natural selection.

Of course, biological features of reality are not the only complex, apparently purposeful features of reality. Many events in the nonbiological realms of science are also difficult to explain on the basis of mere chance, for example, the precise combination of gasses required for biological life, the precise interrelation of laws required for planets and stars to form, the development of biological from nonbiological life (something Darwin's account, because it presupposes biological reproduction, cannot explain), and the development of self-awareness (something apparently unique to humans). If Darwinian selection is not available to explain these phenomena, then we are once again faced with premise 2, a dilemma between chance and an intelligent designer. This has led many philosophers to conclude that, in fact, there are good objective reasons to believe in an intelligent deity.

On the other hand, since natural selection provided a reason to reject teleological conclusions about God for biological life, some philosophers are skeptical of the remaining arguments. Why not think scientists will discover a theory that explains these nonbiological evidences of design? However good the teleological argument may seem for these features, these philosophers argue, the fact that we now know it doesn't work for biological complexity should lead us to suspend judgment about whether it works for nonbiological complexity.

The Ontological Argument

The ontological argument is an evidentialist argument for God's existence that appeals to the very conception of God to show that the nonexistence of God is incoherent or impossible. The word "ontological" comes from two Greek words, *logos*, which we've already seen, and *ontos*, which means "being." So, the word "ontological" means the "study of being." The idea is that, when we think of the idea of "God," we are thinking about the greatest conceivable being—a being so great that no other being can compete with it in perfection. But surely this is not the sort of being that might not exist. If it were, then it wouldn't be the greatest conceivable being—that is, we could think of a greater being, namely, one that could *not exist*. Therefore, the greatest conceivable being is, by definition, one that could not, under any circumstances, not exist. Therefore, God exists.

Anselm of Canterbury (1033–1109) is traditionally considered the first to construct an ontological argument, and he puts it this way, addressing God:

> ...we believe that thou art a being than which nothing greater can be conceived. ... [And the fool] when he hears of this being of which I speak—a being than which nothing greater can be conceived—understands what he hears, and what he understands is in his understanding; although he does not understand it to exist. For, it is one thing for an object to be in the understanding, and another to understand that the object exists. ... And assuredly that, than which nothing greater can be conceived, cannot exist in the understanding alone. For, suppose it exists in the understanding alone: then it can be conceived to exist in reality; which is greater. ... Hence, there is no doubt that there exists a being, than which nothing greater can be conceived, and it exists both in the understanding and in reality (*Proslogium*, Ch. 2).

Anselm argues that when we imagine something, it exists in our minds, and there is a further question as to whether it exists in reality. But the very idea of God is something so great that, to imagine it is to imagine a being that could not possibly not exist—a necessary being. Twentieth-century philosopher Frederick Copleston offers us a simplified version of the argument:

God is that than which no greater can be thought:

But that than which no greater can be thought must exist, not only mentally, in idea, but also extramentally:

Therefore, God exists, not only in idea, mentally, but also extramentally.

The Major Premiss simply gives the idea of God, the idea which a man has of God, even if he denies His existence.

The Minor Premiss is clear, since if that than which no greater can be thought existed only in the mind, it would not be that than which no greater can be thought. A greater could be thought, i.e., a being that existed in extramental reality as well as in idea (*A History of Philosophy*, Vol. II, 1948, p. 162).*

The obvious question is: Why think that *our belief about* a being so great that it must exist implies that this being *actually exists*? A contemporary of Anselm's, Gaunilo (eleventh century), criticizes the argument by offering an absurd parallel: What if I imagine an island, than which there is none greater? If there is none greater, it also should exist both in my understanding and in reality. But it would be absurd to begin searching for this perfect island on the basis of this reasoning. Therefore, it is absurd to believe that God exists on this basis.

But Gaunilo has missed the force of Anselm's argument, and Anselm aptly points this out. Gaunilo takes Anselm to be saying that God is greater than any other being that *actually exists*. But Anselm says that God is a being than which *a greater cannot even be conceived*. Surely, just imagining a greater car or house than actually exists does not imply that such a thing actually exists. But a being greater than that which can be conceived must exist in both our imagination and in reality, otherwise we could imagine a greater being—namely, one that necessarily exists both in our imagination and in reality.

But the argument is not out of the woods yet. Why not imagine an island, than which a greater cannot be *conceived*? Would this, then, imply that such an island exists? Surely, not. But why not? And if not, does the argument for God's existence also fail?

---

*The terminology "major premise" and "minor premise," is an artifice of Aristotelian logic, in which Copleston was trained. The major premise typically referred to a universal claim, such as, "All men are mortal." The minor premise typically referred to a particular claim, "Socrates is a man." And from these two, a conclusion is drawn, "Socrates is mortal." In this case, the major premise expresses the definition of God, which applies everywhere the concept is conceived. The minor premise expresses a particular implication of the major premise—the idea of God entails the existence of God.

Twentieth-century philosopher Alvin Plantinga argues, surprisingly, that this modified version of Gaunilo's objection does fail, but the argument for God's existence does not. Plantinga argues that the idea of a greatest possible island is impossible, whereas the idea of a necessary being like God is not. He writes:

> The idea of an island than which it's not possible that there be a greater is like the idea of a natural number than which it's not possible that there be a greater, or the idea of a line than which none more crooked is possible. There neither is nor could be a greatest possible natural number; indeed, there isn't a greatest actual number, let alone a greatest possible. And the same goes for islands. No matter how great an island is, ... there could always be a greater. ... The qualities that make for greatness in islands—number of palm trees, amount and quality of coconuts, for example—most of these qualities have no intrinsic maximum (*God, Freedom, and Evil*, 1977, 90–91).

Plantinga explains that some properties of an object, particularly quantitative features, have no "intrinsic maximum," that is, there is no greatest one of them, and so no greatest possible one. Just as natural numbers have no intrinsic maximum, neither do the great-making properties of an island. But God, on the other hand, doesn't have quantitative properties. This sort of being has qualitative properties—love, knowledge, power, etc.—of which there can be an intrinsic maximum—or, at least, there is no contradiction in supposing so. Therefore, since a greatest conceivable island is impossible, but a greatest conceivable being is not, the argument does not fail on the basis of our revised Gaunilo-type objection.

Despite these widely-accepted responses to worries about Anselm's argument, there remains one nagging problem. The argument leads to some important conclusions: A maximally great being would necessarily exist, and if a maximally great being is even possible, then it actually exists—because it is *maximally* great (not just sort-of great). But the question remains is a maximally great being possible? Plantinga thinks so, but admits that this is no proof of God's existence, since someone else might disagree. If there is no argument for the claim, "a maximally great being is possible," then the argument hits a wall. Some will think so, and conclude that the argument is sound; others will not think so, and conclude that the argument is unsound. And whichever you choose, it is not obviously a function of reason, which means it doesn't count as a sufficient argument for God's existence.

The Cosmological Argument

The cosmological argument is an evidentialist argument that appeals to the origins of the cosmos as evidence that God exists. "Cosmological" comes from two Greek words, *logos*, and *kosmos*, which means "universe." The basic idea is that there must be a reason why there is something rather than nothing. Natural laws cannot explain why there is something rather than nothing, since natural laws are included among the things that are something; therefore, they also need an explanation. In addition, stuff just doesn't pop into existence out of nothing. There is a very old, widely-accepted dictum that says, *ex nihilo, nihil fit*, that is, "out of nothing, nothing comes." Therefore, if law doesn't explain why there is something rather than nothing, and pure random chance doesn't explain it, then a being that can create and intend must be responsible.

One example of a cosmological argument is found in Thomas Aquinas's Five Ways. Thomas offers five arguments for God's existence that have been unfortunately characterized as "proofs." A *proof* is a logical or mathematical deduction, and while parts of Thomas's arguments are deductive, the final step of each is inductive, and Thomas does not try to hide this. After each argument, he concludes, "And this is what we call God." The reason for this inductive move is simple. The Christian conception of God, which Thomas defends, is much richer than anything he can prove in an argument—including properties like aseity, immutability, omniscience, omnipotence, creator, designer, etc. So, when Thomas offers an argument for the claim that there must have been a First Mover for any other movement to occur, he points to the list of properties of God and says: that's on the Christian's list. So, the idea is that the more properties he can prove, the more likely it is (inductively) that the Christian God exists.

Thomas's Second Way is a cosmological argument, which he explains like this:

> The second way is from the nature of the efficient cause. In the world of sense we find there is an order of efficient causes. There is no case known (neither is it, indeed, possible) in which a thing is found to be the efficient cause of itself; for so it would be prior to itself, which is impossible. Now in efficient causes it is not possible to go on to infinity, because in all efficient causes following in order, the first is the cause of the intermediate cause, and the intermediate is the cause of the ultimate cause, whether the intermediate cause be several, or only one. Now to take away the cause is to take away the effect. Therefore, if there be no first cause among efficient causes, there will be no ultimate, nor any intermediate cause. But if in efficient causes it is possible to go on to infinity, there will be no first efficient cause, neither will there be an ultimate effect, nor any intermediate efficient causes; all of which is plainly false. Therefore it is necessary to admit a first efficient cause, to which everyone gives the name of God (*Summa Theologica*, Question 2, Article 3, trans. Fathers of the English Dominican Province, 2008, Kevin Knight).

There is a lot here, so we will need to examine it in pieces. Thomas's first premise is that:

1. **In the universe, everything has a cause.**

This he derives purely from experience. Have you ever experienced anything that exists that wasn't caused by something else? Neither have we. So, now the question becomes: Where did all this stuff come from? So, premise 2 eliminates one option:

2. **Nothing can be the cause of itself.**

Why should we buy this premise? Thomas tells us that it leads to contradiction:

  a. **If something causes itself, then it must exist prior to itself.**

  b. **Nothing exists prior to itself.**

  c. **Therefore, nothing can be the cause of itself.**

If 2 were true, something would have to exist to bring itself into existence. But this means that something would both exist and not exist at the same time, which is impossible. If nothing causes itself, then there are only two options left, which is premise 3:

3. **Either there are infinite causal chains or there is a first uncaused cause.**

The idea, here, is that the universe is either infinitely old or it was caused by something that doesn't require a cause. Thomas repeats an old argument from Aristotle for the conclusion that:

4. **There can be no infinite causal chains.**

What is the old argument for this claim? Thomas gives only the briefest sketch of it, but the idea is that infinite chains, by definition, have no first member. But causal chains cannot be conceived without first members; if there are no first members, there are no successive members. Therefore, if there were no first member of a causal chain, that is, if it were infinite, then there would not be any events now (which are successive events). Of course, there are events now; thus, the causal chain that produced the events now is not infinitely long. Thomas sketches it this way:

  a. **If a chain of causes were infinite, then there would be no first cause.**

  b. **If there is no first cause, there can be no subsequent causes or causes "now."**

  c. **There are causes now.**

  d. **Therefore, the chain of causes cannot be infinite.**

Now, since nothing can cause itself, and since the causal chain leading to the present cannot be infinitely long, the only option left is a first, uncaused cause:

5. Therefore, there must be a first uncaused cause.

And this is one of the attributes of the traditional Christian God, so Thomas notes:

6. This first uncaused cause is what we call "God."

There is no doubt that this argument is clever and intuitive. However, before we accept it as sufficient for establishing the existence of God, we must be able to respond to three common worries with it.

First, even if Thomas establishes a first, uncaused cause for every causal chain, it is not clear that there is only one causal chain. There may be a causal chain leading from the beginning of the world to Jamie, and there may be another leading to Hitler. Since Thomas's argument does not establish that there is only one causal chain, it is possible that there are multiple first, uncaused causes (see Fig. 9-2).

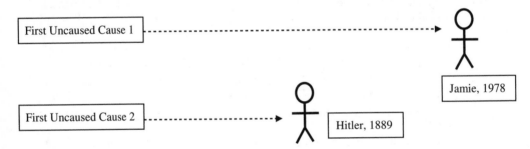

FIGURE 9-2 · Multiple uncaused causes.

But this is certainly nothing like the "God" that Thomas is attempting to prove exists. Therefore, if this argument is to be successful, someone must restrict the possible number of causal chains to one.

Second, Thomas derives his first premise from experience of objects in the universe. But it is possible that the universe as a whole is very different from the objects inside it. To conclude that a set of objects must have a property of its members is to commit a logical fallacy known as the *genetic fallacy*, or the *fallacy of composition*.

For example, imagine that Rob runs a famous school of music, and the musicians who graduate are widely regarded as some of the most exquisite in the world. Now imagine that you discover an exquisite and famous orchestra comprised only of students from Rob's school. Could you conclude that, since Rob is responsible for the exquisiteness of each musician, he is also responsible for the exquisiteness of the whole orchestra? Anyone who plays an instrument knows the answer is no. A person can be a wonderful musician, but be unable to "click" with just anyone. For a band or orchestra to be good, they must have a certain connection with one another. Rob may be responsible for hand-picking this orchestra, or someone else may be. Or perhaps they got together themselves. The point is that, just because Rob is responsible for the individual members, he is not necessarily responsible for the whole.

This worry translates directly to Thomas's argument. Just because all of the events he has experienced have a cause, it does not follow that the causal chain itself, or the universe as a whole, has a cause.

And third, both Aristotle and Thomas were operating under an understanding of "infinity" that was overturned in the nineteenth century. Both Aristotle and Thomas concluded that, because an infinite set has no first member, it can have no succeeding members. However, mathematicians in the 1800s showed that the concept of an actually infinite set is not incoherent and does not necessarily lead to contradiction. Therefore, just because there is no first cause of a causal chain, it does not follow that each effect does not have a cause. For every effect, there is a cause; and there may be an infinite number of causes and effects. Therefore, anyone wanting to defend a cosmological argument must either resolve these difficulties or pursue an alternative line of reasoning.

Are there any plausible prospects? Around the same time that Thomas was writing, an Islamic philosopher named al-Ghazali was also using Aristotle to defend the existence of God. His argument is called the "Kalam Cosmological Argument." The word *kalam* is the Arabic word for "nature." The interesting feature of al-Ghazali's argument is that it is not subject to all of the same criticisms as Thomas's. One clear way to formulate the argument is offered by William Lane Craig (*The Kalam Cosmological Argument*, Eugene, OR: Wipf and Stock, 1979):

1. Everything that begins to exist requires a cause for its origin.

This premise avoids the genetic fallacy because it refers only to things that begin to exist. If the universe didn't begin to exist, we could just focus on the events in the universe, and derive a first cause from them. In addition, there may be a number of objects that did not begin to exist, e.g., abstract objects like numbers, propositions, etc. Like an uncaused cause, these would not require an explanation in terms of causes.

In addition to premise 1, al-Ghazali argues that there are good reasons to believe that the universe began to exist:

2. The world began to exist.

How does he justify this premise? In its original form, al-Ghazali points to the same Aristotelian argument from infinity that Thomas does:

a. **There are temporal phenomena in the world.**

b. **These are preceded by other temporal phenomena.**

c. **The series of temporal phenomena cannot regress infinitely.**

    c.1. **Why? An actually existing infinite series involves various absurdities.**

d. **Therefore, the series of temporal phenomena must have had a beginning.**

As we have seen, this argument has the defect that an actually infinite set may not involve the absurdities Aristotle and Thomas think they do. But we can revise both Thomas's and al-Ghazali's arguments in light of contemporary mathematics by replacing c.1 with:

    c.2. **An actually infinite set cannot be completed by finite addition.**

This claim is consistent with contemporary mathematics of infinity. For any set containing members of a finite size, either that set contains an infinite number of members or a finite number of members. If it is finite, no amount of finite members can be added to make the set infinite. If it is infinite, then all the members came to be simultaneously. Therefore, consider all of the past moments of time leading up to the present. Either (A) the past set is infinite or (B) it is finite (see Fig. 9-3).

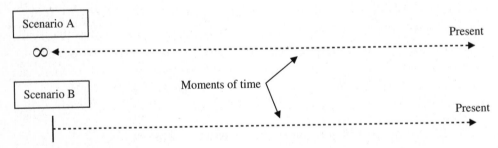

FIGURE 9-3 · Either the past set of moments is infinite or it is finite.

If it is infinite, as in Scenario A, there can be no time at which it was not infinite (because then an infinite set would have been constructed by adding in finite succession, which is mathematically impossible). If there is no time at which it was not infinite, then it either existed as a completed infinite set from eternity, or it was brought into existence as a completed infinite set. If it existed as a completed infinite set from eternity, we merely change the question; we are interested in how that set came to produce the causal chain that leads to the present. It is apparently the uncaused cause we are after. Alternatively, if the set was brought into existence as a completed infinite set, then we are back to al-Ghazali's premise (c): The series of temporal phenomena cannot regress infinitely. From this discussion, al-Ghazali's conclusion follows:

3. Therefore, the world has a cause for its origin: its Creator.

So far, so good. We have constructed a cosmological argument that avoids all the problems that Thomas's version faced. It does not commit the genetic fallacy, it is not committed to an outdated view of infinity, and it does not allow the possibility of many causal chains because the past events that cannot be infinite are the moments of time, not particular causal chains within time.

Still Struggling

This argument relies heavily on the idea of "infinity." The concept of infinity has plagued philosophy from its earliest days. It can be defined in many ways (e.g., qualitatively, quantitiatively, as sets of sets, as sets of points) and it leads to many puzzles (e.g., Zeno's paradoxes of the runner and the arrow, Hilbert's hotel, the continuum hypothesis). Nevertheless, many of us have an intuitive grasp of the infinite in our understanding of the integer line. The integer line consists of a series of whole numbers (no fractions or decimals) that continue infinitely in both directions:

$$\leftarrow \ldots -10\ -9\ -8\ -7\ -6\ -5\ -4\ -3\ -2\ -1\ \mathbf{0}\ 1\ 2\ 3\ 4\ 5\ 6\ 7\ 8\ 9\ 10 \ldots \rightarrow$$

What do we mean by "infinite" in this case? Simply this: for any particular integer, there is one smaller and one larger—no matter how small or large an integer you choose.

How "infinity" helps the Kalam:
Now, the Kalam cosmological argument takes it for granted that you perceive time progressing forward from the past to the future (this may or may not be true, but nonetheless, you perceive it this way). The present (the "now") is distinguished from the past by the difference in events at each time (it was once true that Julius Caesar was alive, now it is not). These events can be numbered using integers, and all events were caused by events prior to them (Julius Ceasar's ceasing to be alive was caused by the senators killing him). Likewise, your perception right now that you are reading a book was caused by a series of prior

events that include purchasing or borrowing this book). Your reading this book is the last member of a series of events leading to this moment. The question is: could that series of events extend infinitely into the past?

Proponents of the Kalam say no. This is because there are only two options for producing a new infinite series: either begin with an infinite series (say, series X) and add to it (so it becomes series X*), or begin with a finite series (say, series Y) and add to it until it is infinitely large (Y + a + b + c + d…). The latter, however, is impossible. For every event you add, you still have a finite series. It is impossible to construct an infinite series this way because of the definition of infinity we began with: for every finite event you add, you can add another; and a finite addition to a finite set only ever produces another finite series (finite series "Y + a"; finite series "Y + a + b"; finite series "Y + a + b + c," and so on).

What about the former option: begin with an infinite series and add to it? This is more plausible, but now the issue is slightly different. Now the question is: how does your perception of the events "now" (reading this book) come to exist (since you perceive that they do)? The perception of the flow of time can't be a part of the infinite series to which finite events are added (since that series is already complete and there is not a time in that series prior to which any perception in that series exists); therefore, it can only be a part of the finite series of events that are added to the original infinite series to make a new infinite series. This means that it is possible for the past to consist of an infinite series of events, but not in a way that helps those who disagree with the Kalam. This is because of the first premise: everything that begins to exist has a cause. Either the original infinite series as a whole had a cause, in which case proponents of the Kalam win, or the original infinite series is uncaused, in which case, something apart from the uncaused infinite series must explain the "beginning to exist" of those caused events (those finite events added on, namely, the perceptions of the flow of time), in which case, proponents of the Kalam win.

So, do we have an evidentialist argument for God's existence that works? Most philosophers agree that the Kalam argument is the strongest candidate. There is one worry that might upset the argument, but we will leave it to you to decide whether it is sufficient.

In the development of quantum physics, scientists discovered that some sub-atomic particles (i.e., particles smaller than atoms: electrons, muons, gluons, tachyons, etc.) seem to exhibit uncaused behavior. Radioactive decay is a good example. Radioactive particles decay spontaneously by half of their members at a steady rate. However, we cannot predict which particles will disintegrate at any point in the decay cycle. Thus, it would appear that the choice among decaying particle is uncaused.

A *quantum fluctuation* is perhaps the most interesting of these events. In a quantum fluctuation, there is a temporary change in the energy at a point in space. Particles appear to pop into existence uncaused. Although we can know roughly when and where they will pop in and out of existence, there is no discernable causal explanation for this behavior. If this is right, it wreaks havoc on premise 1 of the Kalam cosmological argument—that everything that begins to exist has a cause.

To be sure, this is not to say that the universe arose spontaneously from a quantum fluctuation. This is, by definition, impossible, since the beginning of the universe explains the origin of quantum physical laws, not vice versa. But if it is even possible that something begins to exist uncaused, premise 1 is false.

Is the Kalam done for? That depends. There are two available interpretation of these quantum fluctuations. The first interpretation is the one we've already noted, namely, that there are uncaused events. This is known as the *Copenhagen interpretation* of quantum behavior, and proposes that quantum physics disproves classical mechanics (which presupposes that events have causes). A second interpretation was developed by Louis de Broglie (1927) and David Bohm (1952), according to which, the causal factors involved in quantum events are simply not available to us—they are "hidden variables." This view is known as *Bohmian mechanics* or the *Broglie–Bohm* interpretation of quantum behavior.

Broglie and Bohm offered explanations of the quantum events that do not require giving up the classical picture of mechanics. There is widespread controversy over which interpretation should be accepted as the most plausible. However, if the proponent of the Kalam argument has available a physical explanation of quantum events that is consistent with classical mechanics and the intuitive notion that "out of nothing, nothing comes," the Kalam remains one of the more plausible arguments for God's existence.

A Brief Note on Prudential Arguments

Prudential arguments make up a unique category of arguments for the existence of God. Whereas evidentialist arguments attempt to justify belief that the claim, "God exists," is true or false, prudential arguments attempt to establish that you should believe (or not) *regardless of whether* the claim, "God exists," is true.

What would such an argument look like? We have some interesting examples from the history of psychology. For instance, a plethora of research suggests that terminally ill patients who are optimistic about their chances of survival, despite their doctor's negative prognosis, have a much better overall quality of life than patients who are realistic about their prospects. This irrational belief in spite of the evidence has a number of practical benefits—including higher spiritual well-being, lower anxiety, and lower depression. Therefore, it would seem it is in a terminally ill patient's best interests to believe irrationally in order to obtain the practical benefits. This is a prudential reason for believing something you have evidence is not true.

Prudential arguments for God's existence work similarly. They suggest that believing in God is in your best interests even if you have no evidence at all as to whether God exists. Blaise Pascal (1623–1662), a French philosopher, mathematician, and co-inventor of the probability calculus, offered a well-known prudential argument for God's existence, often called the *Wager Argument*.

Pascal reasoned that: If you believe in God and you're wrong, you lose very little—perhaps you don't pursue quite as many carnal pleasures. And if you believe in God and you're right, you gain infinite happiness in heaven with God after you die. On the other hand, if you don't believe in God and you're right, you gain very little—you can pursue a few more of those carnal pleasures. But if you don't believe in God and you're wrong, you lose an infinite amount of happiness in hell, separated from God. Now, just do the math: Since infinite gains always offset finite losses, and infinite losses can never be offset by finite gains, it is in your best interests to believe in God despite the evidence.

Some object to Pascal's Wager, arguing that Pascal's argument doesn't tell you which god to bet on, and if you get it wrong, you still face infinite losses. While this is true, prior to offering the Wager, Pascal has offered quite a bit of rational evidence that Christianity is the only plausible religious choice. But since reason can take you no further than this, Pascal then offers the prudential reason for believing in the Christian God over not believing at all.

There are a number of interesting prudential arguments, including a famous argument from William James in an article called, "The Will to Believe." And many include some sophisticated reasoning with probabilities. Given space constraints, we will move on to an evidentialist argument against God's existence, but consider the suggested readings at the end of this book for more on prudential arguments.

An Argument against God's Existence: The Problem of Evil

The problem of evil is an evidentialist argument against the existence of God. It has two versions, one deductive, which is known as the "logical problem of evil," and one inductive, which is known as the "evidential problem of evil." Though this argument has been called the problem of "evil" for a long time, it is probably better named the problem of "pain" or "suffering," since the force of the argument does not depend on any substantive philosophical conception of "evil." Therefore, in our discussion, we will call them the logical and evidential problems of *suffering*.

The Logical Problem of Suffering

The logical problem of suffering is a deductive argument, which, if sound, rules out even the possibility of the traditional conception of God. The idea is that the traditional conception of God involves claims that are logically incompatible with the existence of evil. David Hume put it like this: "• Is [God] willing to prevent evil, but not able? then he is impotent. • Is he able, but not willing? then he is malevolent. • Is he both able and willing? then where does evil come from?" (*Dialogues Concerning Natural Religion*, Part 10, p. 44, edited by Jonathan Bennett).

We can formulate this argument more formally as follows. The traditional conception of God is that:

1. **God is all-powerful, all-knowing, and all-good.**

From this conception, so the argument goes, we can formulate the following three claims:

2. **If God is all-powerful, he can rid the world of any suffering he knows about.**

3. **If God is all-knowing, then he would know about any suffering that exists or is about to exist.**

4. **If God is all-good, then he would want to rid the world of any suffering he can and that he knows about.**

If 1–3 are true, it necessarily follows that no suffering exits. God is just not the sort of being who would allow any suffering to exist. Unfortunately for theists, it is difficult to not also accept the claim:

5. **At least some suffering exists.**

But if 5 is true, then the antecedent of one of 2–4 must be false (that is, either God is not all-powerful, not all-knowing, or not all-good). Yet, if any of these three antecedents is false, premise 1 is false, and a being like the traditional conception of God does not exist. Given premise 5, the argument concludes:

6. **Therefore, premise 1 is false, and the traditional God does not exist.**

This argument is clearly deductively valid, so if a theist wishes to challenge its soundness, she must focus on the truth of the premises. Giving up 1 is not an option if she is a traditional theist. But 2–3 are also very intuitive. Nevertheless, a few philosophers have highlighted a problem with premise 4.

If it is even possible that the traditional God has some reason to permit suffering, then premise 4 is false. What reason might be good enough to allow suffering? Aurelius Augustine (354–430 CE) suggested that, perhaps some goods cannot be realized without suffering. For instance, the virtues of charity, generosity, grace, and forgiveness could not be realized unless two conditions were met: (1) creatures are free to act morally or immorally, and (2) some sort of suffering is possible that a creature could react virtuously toward. Augustine argued that it is a greater good to experience generosity than it is an evil to experience lack. This response is known as the "Greater Goods Defense" response to the logical problem of suffering.

The Greater Goods response actually includes another response that is sufficient for undermining premise 4 on its own, that is, that humans are free to act morally or immorally. If God has any reason whatsoever to permit actual, honest-to-goodness free will, then he cannot, at the same time, prevent the possibility of immoral suffering—it is logically impossible.

Some may worry at this point: if we admit this, haven't we already given up divine omnipotence? The answer, according to most theists, is no. It is no limitation on God's power that he cannot bring about logical impossibilities; in fact, it is a testament to his greatness. Logical impossibilities are absurdities, like round squares, married bachelors, and 2 + 2 = 5. If God could make a logical impossibility true, it would also entail that he both exists and doesn't exist at the same time—more absurdities. Therefore, God cannot make rocks too big for him to lift, cannot make non-self-identical objects, and cannot create truly morally free beings without allowing the possibility of immoral suffering.

Are there any good reasons God might have for giving creatures free will? We can think of a few (and Augustine's is as good as any), but it is not necessary. As long as it is not contradictory to imagine God giving creatures free will—that is, as long as it is possible that he has a reason—premise 4 is false. This is all that is required to show that there is no logical contradiction in the existence of God (premise 1) and the existence of suffering (premise 5). This is called the "Free Will Defense" response to the logical problem of suffering.

The Evidential Problem of Suffering

By far, the most worrisome argument against the existence of God is the Evidential Problem of Suffering. The Evidential Problem is inductive, rather than deductive. It attempts to show that either the amount or the quality of suffering in the world makes it highly unlikely that a being like the traditional God exists.

In light of the response to the Logical Problem of Suffering, both the theist and the atheist can agree that suffering exists, that some suffering is consistent with the existence of God, and that some of the suffering that exists is really, really bad—unimaginable in some cases. In addition, both the theist and the atheist can agree that there is an amount of suffering that is so great that it would be impossible for God to allow. This would have to be some suffering that was truly (not merely from a human perspective, but really) random or meaningless, or so extensive that not even God could use it to bring about a good that would justify it—suffering that couldn't be used for anything, even by God, is called "gratuitous" suffering.

So, both theist and atheist agree on the first premise of this argument:

1. If God exists, there will be no gratuitous suffering in the world.

The atheist then argues that:

2. It is highly likely that there is gratuitous suffering in the world.

And concludes that:

3. Therefore, it is highly likely that God does not exist.

Since the theist accepts premise 1, the burden on the atheist is to defend premise 2. And many atheists think this is a slam dunk. Just read any of the biographies of Holocaust survivors; the accounts of the tens of millions of people that Russian communist leader Stalin killed; the millions of people that Cambodian communist leader Pol Pot killed; the crusades, which led to the deaths of innocent people because of a disagreement over religion; the innocent lives ruined by the atomic bomb; the fire-bombing of civilian targets in Dresden, Germany during World War II; the 200,000 Iraqi citizens that Saddam Hussein gassed to death; and we could continue on indefinitely. Are these not examples of gratuitous suffering?

The theist is not convinced, and for good reason. At what point does suffering become gratuitous, and who gets to decide? Rob thinks that getting on an airplane is gratuitous suffering, and Jamie is willing to consider the possibility that watching football is a means of gratuitous torture. And we both recognize that these are infinitesimal compared to the suffering of the Holocaust. But are our preferences the relevant standard of evaluation? We hope not. The theist can admit that all the events in the list above are bad, even *really* bad. But, if God exists, then, by definition, they are not gratuitous.

When does "really bad" become "too much bad"? Imagine that someone brings you a gallon of milk and asks: "Is this too much?" Before you can answer you have to ask: "Too much for what?" With suffering, we don't have the luxury of asking, "Is this more suffering than God can use?" And if the theist hasn't been let in on the secret, the atheist probably hasn't been, either. Therefore, it is not clear how the atheist would establish, beyond brute intuition about the "badness" of some suffering, that there is more suffering in the world than even God could use.

Therefore, if a person has more reason to believe that God exists than that there is gratuitous suffering in the world, the evidential problem of suffering does not go through. On the other hand, if a person has more reason to believe that there is gratuitous suffering than that God exists, the evidential problem of suffering goes through.

So, at the end of this chapter on philosophy of religion, we leave you, the reader, with the following decision. Is it more likely that:

**2.a. It is highly likely that there   or that   2.b. It is highly likely that
is gratuitous suffering in   God exists.
the world.**

Which you choose will determine which of the following conclusions you are committed to:

**3.a. Therefore, it is highly like   or   3.b. It is highly likely that is no
that God does not exist.   gratuitous suffering.**

QUIZ

1. According to what we said in this chapter, the following are all questions that philosophers of religion ask except:

    A. Why believe gods or a God exists?

    B. What sort of being must a god be like?

    C. What religious text must one believe in order to have eternal life?

    D. How reliable is a particular sacred text?

2. The most common argument against the existence of God is that

    A. there need not be a first cause of everything that exists.

    B. there is too much evil or suffering in the world to think that such a being is likely to exist.

    C. there is no way to prove the existence of God.

    D. None of the above.

3. Non-evidentialism is the view that belief or disbelief in God requires objective evidence.

    A. True

    B. False

4. _____ is the view that evidentialism is true and that there is no objective evidence sufficient for believing either that God exists or that God doesn't exist.

    A. Fideism

    B. Evidentialism

    C. Agnosticism

    D. Prudentialism

5. _____ argued that people can believe in God only if they have experienced the "internal instigation of the Holy Spirit," which is a subjective experience.

    A. John Calvin

    B. John Locke

    C. William James

    D. William Calvin

6. Someone who offers a deductive proof (_____) for or against God's existence faces a much more difficult task than someone who offers an inductively forceful argument (_____).

    A. a cogent argument, a sound argument

    B. a cogent argument, a valid argument

    C. a cogent argument, a strong argument

    D. a sound argument, a cogent argument

7. The idea of a First Mover is central in the

    A. teleological argument.
    B. cosmological argument.
    C. ontological argument.
    D. Watchmaker argument.

8. Newtonian physics presents a challenge to the Kalam cosmological argument.

    A. True
    B. False

9. The logical problem of suffering is a deductive argument, which, if sound, rules out even the possibility of the traditional conception of God. The idea is that the traditional conception of God involves claims that are _____ incompatible with the existence of evil.

    A. empirically
    B. evidentially
    C. logically
    D. None of the above

10. The cosmological argument attempts to show that either the amount or the quality of suffering in the world makes it highly unlikely that a being like the traditional God exists.

    A. True
    B. False

Moral Philosophy

Can morality be discussed rationally? Is there some way to test whether an act is right or wrong? In this chapter, we will introduce the basics of moral philosophy, which, for the most part, attempts to show how both of these questions can be answered with a resounding "yes." We will explain how the study of morality is organized by philosophers, the concept of "normativity," and the three most prominent accounts of what makes an act right or wrong.

CHAPTER OBJECTIVES

In this chapter, you'll learn about...

- The nature of moral philosophy
- How a moral decision is different from any other decision
- Normativity
- Duty-based moral theories
- Consequence-based moral theories
- Character-based moral theories

Should you help a stranded motorist on the side of the road? Should you tell your boss that your co-worker has been stealing office supplies regularly for the past year? Why wouldn't you tell your best friend's wife that your best friend has been cheating on her? Should you abort an unwanted fetus? Further, what grounds or reasons or justifying principles are you appealing to when you face these decisions? In other words, *why* did you lie to your friend that time? If you are a doctor, *what grounds might you have* for treating a police officer rather than a homeless man? *Which principle are you appealing to* when you claim that you disagree (or agree) with capital punishment? Are there objective moral rules that apply in all types of situations, no matter what country or culture you're living in?

These are the kinds of questions that people who study *moral philosophy* ask. In essence, moral philosophy (also known as *ethics**) is the branch of philosophy that investigates and critiques:

(a) **human actions that affect morally relevant beings (definitely humans; probably many animals) and**

(b) **the principles that people appeal to when they act.**

Most moral philosophers identify rules or principles people should follow when making a decision, some of which seem trivially true—like "Murder, rape, and theft are immoral" and "Needless harm is immoral"—and some of which are more controversial—like "Abortion is immoral, even if the mother's life is in danger" or "It is immoral to have more money than you need." These principles help inform our decisions. But these principles need an explanation. For this, moral philosophers turn to moral theories. In this chapter, we'll take a look at what constitutes moral philosophy and three traditional moral theories.

---

*In the past hundred years or so, the term "ethics" has become conflated with "morality." Prior to this, "ethics" just meant the study of "practice" or "habit." So, your "business ethic," would just be how you go about the work of business. You might have heard the term, "Protestant Work Ethic." This typically refers to working really hard for long hours, and it does not refer to a moral features or implications of an act; it merely refers to how people go about working. For our purposes, "ethics" and "morality" will refer only to the moral features or implications of an action. The study of "habit" or "practice" is now called "axiology."

What Is "Moral Philosophy"?

As you can see from the figure below (Fig. 10-1), moral philosophy is a branch of philosophy:

PHILOSOPHY:
The systematic study of reality using good reasoning in order to clarify difficult questions, solve significant problems, and enrich human lives. There are five major branches of philosophy, and those branches themselves have sub-branches.

METAPHYSICS:	EPISTEMOLOGY:	LOGIC:	POLITICAL PHILOSOPHY:	MORAL PHILOSOPHY:
The study of the nature of existence, including the kinds of things that exist.	The study of knowledge, including how we come to know things, the justification for knowledge, and type of truth.	The study of the principles, of correct reasoning, including argument identification, formation, and analysis.	The study of the justification for organizing human interaction in social settings which includes defining and justifying rights, laws, justice, and appropriate forms of government.	The study of human actions that affect beings capable of being harmed in some way (definitely humans, as well as many animal species) and the principles that people appeal to when they act.

FIGURE 10-1• The branches of philosophy.

Moral philosophy also has sub-branches that include metaethics, normative ethics, and applied ethics, as can be seen in the diagram below (Fig. 10-2). In this chapter, we will focus on normative ethics, and in the following chapter, we will discuss some issues in applied ethics.

MORAL PHILOSOPHY/ETHICS:
The study of human actions that affect beings capable of being harmed in some way (definitely humans, as well as many animal species) and the principles that people appeal to when they act.

METAETHICS:	NORMATIVE ETHICS:	APPLIED ETHICS:
• nature of moral knowledge • proper grounds for justifying moral claims • metaphysical/ontological status of moral norms and entities	• what constitutes the "good life?" • what should I do, and who should I be? • development, analysis, and critique of various moral/ethical theories	• realm of professions, institutions and public policy • generates practical moral answers • applies moral/ethical theories to practice

FIGURE 10-2• The branches of ethics.

Business ethics, legal ethics, environmental ethics, cyber-ethics, bioethics, and others are sub-branches under applied ethics (Fig. 10-3). Applied ethics plays an important in any society since, for example, virtually all of the topics in bioethics have been, at one time or another, "hot button" issues in politics in the United States and many other parts of the world.

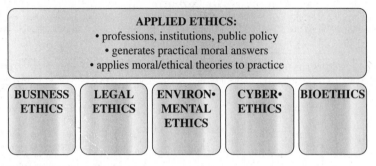

FIGURE 10-3 · Some branches of applied ethics.

How Is a Moral Decision Different from Any Other Decision?

We make all kinds of decisions daily: to brush our teeth, to go to the store, to eat a fourth piece of cake, to watch TV, to scream at our kids, to exercise, to cheat on our taxes. But which decisions are actually *moral* ones, that is, which decisions have moral implications? Actually, all decisions have a moral dimension. Since decisions are a type of action, we will refer to actions in the rest of the chapter. For any act, X, X is

1. Permissible,
2. Impermissible,
3. Obligatory, or
4. Supererogatory.

These categories identify the *moral implications* of an act. Moral implications are the facts that change given your decisions. For example, if you knowingly choose to steal from your little sister, the moral implication is that you are now *blameworthy* for committing an immoral act, and you may *deserve* punishment, or at least, you may *owe* your sister restitution. The italicized words in the preceding sentence are the "moral implications" of your act. They are negative moral implications because you have violated a moral obligation. Positive implications include praiseworthiness or virtuousness.

A *permissible* act is one from which no positive or negative moral implications ensue. For example, picking up your coffee mug, tapping your foot, etc.

Impermissible actions are those that violate an obligation, that is, some rule that has determined the action to be immoral. For example, rape, murder, etc. Obligatory actions are those that it would be wrong not to perform. For example, refusing to give money back that doesn't belong to you, telling the truth. And supererogatory actions are permissible actions that are morally good, but not obligatory; you can be praised for them even though you didn't have to do them. For example, charity work or "jumping" a stranger's dead battery.

So, every action has moral implications; that is, it will fall under one of the four categories above. And the morality of an action is expressed in a special kind of claim, called a "normative" or "prescriptive" claim.

Normativity

In addition to falling into one of the above categories, moral claims are *normative*, which means there is some obligation associated with them, regardless of whether we acknowledge it. When someone says, "Do no harm," "You should not murder," or "You ought to tell the truth," these are moral claims intended to apprise a person of her obligation to act a certain way. Normative claims are contrasted with *descriptive* claims, which merely describe what is the case concerning a thing, object, event, or phenomenon (see Fig. 10-4).

Descriptive Claim:	Moral Prescriptive Claim:
Communicates what *is* the case, or *describe* a thing, event, object, or phenomenon in reality.	Communicates what *ought to be* or *should be* the case, or *prescribe* what one ought to do in a moral situation.
EXAMPLES: • It is sunny out right now. • That pasta is undercooked. • Most people eat in the U.S.A. eat more than 2,000 calories a day. • I am feeling tired. • Oklahoma is a state in the U.S.A.	EXAMPLES: • One ought not commit fraud. • We should not allow grade inflation to occur in our schools. • Humans should not kill animals. • You should reward effort as well as merit.

FIGURE 10-4 · Descriptive vs. prescriptive claims.

BOX 10-1 Types of Prescription

There are *moral prescriptive claims* and there are *non-moral prescriptive claims*. Moral prescriptive claims are those where some being is capable of being harmed, whereas non-moral prescriptive claims do not have this quality, but still tell you what you ought to do in some situation.

Examples of moral prescriptive claims: Don't kill needlessly; You should not steal ever; You should seek the truth of the matter before prosecuting or punishing someone; You ought to tell the patient the whole truth about his situation.

Examples of non-moral prescriptive claims: You ought to draw your cursive letters this way, Johnny; You should not hold the violin between your knees to play it, Sally; You should go to a state college rather than a private college, Jim; You should save your money for a new car, Fred.

Rationality

The truth of a normative claim is determined in much the same way as a descriptive claim, that is, by rational, rather than emotional considerations. This means that principles of logic and critical thinking should be applied to objective evidence in order to justify a decision. Reason assists in "leveling the playing field," so to speak, so that moral decisions can be both *consistent* and *fair*. For example:

If we appeal to our own personal feelings, wants, desires, or inclinations, rather than reason, then our moral decisions will be subjective, and hence, inconsistent. Frank could say, "I need to steal from the grocery store because it benefits me. Period." Well, what about the rest of us who shop there? What about those folks who worked so hard to produce the goods and therefore, own them? What about those who obtained those good fairly by trading money for them? The emotional interests of the shopper and the owners are in conflict. Without reason, one cannot determine who is morally justified in his actions. One can't resolve conflicts among feelings, wants, desires, or inclinations without reason.

Similarly, if we appeal to our own religious preferences, rather than reason, our moral decisions will be subjective, and hence, inconsistent. As with the grocery case, this could lead to harm to innocent people, as has been the case throughout human history when one oppressive group or tyrant has appealed to a god as a justification for very harmful decisions made. The Gods of Christianity and Islam have been used to justify torture and killing. Hitler invoked the Christian God in his plan, too. The terrorist attacks of 9/11 were claimed to be motivated by the will of Allah. The religious interests of Christianity and Islam are in conflict. Without reason, one cannot determine which, if either, group is justified in its actions. One can't resolve conflicts among religious claims without reason.

In both of these paragraphs, we have offered examples of how rejecting reason can lead to inconsistency. Inconsistency undermines meaningful discourse. Without meaningful discourse, any investigation into reality is doomed.

To summarize so far, every action is either permissible, impermissible, obligatory, or supererogatory, all moral claims are normative, in some sense, that is, that they refer to the moral implications of an act, and the truth of moral claim is determined primarily on rational considerations. The question then becomes: *How do we tell which actions fall into which of the four categories?* To answer this question, we must turn to moral theories.

Moral Theories

Moral theories explain what makes an act permissible, impermissible, obligatory, or supererogatory. Each theory applies a criterion or set of criteria to an action at a time and renders a judgment on the moral status of that action. In this sense a moral theory is a *decision theory*, that is, it is a procedure for deciding how you should act in a set of circumstances. Different moral theories convey different truth values for the same action, so it is important that we carefully evaluate each so that we adopt only the most plausible decision theory. We will consider three types of classic moral theory.

Duty-Based Theories

The first type, which we will call cuty-based theories, is based on the work of the German philosopher, Immanuel Kant (1724–1804). Kant's moral theory sets him apart as the foremost proponent of *deontology*. The term is a combination of two Greek words: *deontos*, which means, "duty," and *logos*, which means, "the logic of" or "the study of." Kant argues that whether an act is moral or immoral depends wholly on one's *duty* to act according to a moral principle concerning that action.

Duties, for Kant, are not a matter of considering the consequences of an action. They are a matter of pure rationality. If you understand only the facts of a situation, Kant argues, you can derive your duty in that situation. Duty-based theories stand in contrast to consequence-based theories. According to consequence-based theories, whether an act is moral or immoral depends wholly on the consequences of the action. Once you understand the likely consequences of your act, you will understand your moral obligations.

Kant articulates his duty-based theory in three forms, two of which we'll mention here. The first can be paraphrased as follows:

> Whenever you act, make sure that your action is something that can be universalized without contradiction. In other words, ask yourself the question: "What if everyone did what I'm about to do?" and if it undermines or negates what you want to do, then it's immoral and you should not do it.

For example, let's say you wanted to borrow money from someone knowing that you will not pay it back. Now, think what would happen logically if all people *universally* did this: the very idea of "borrowing" would completely go away, since no one would ever trust another person to borrow because Person A would know that she/he would never get the money back from Person B.

And, you yourself, then, could never borrow any money with *or without* the intent of paying back! So, you would be contradicting or undermining your own action, which is irrational and unreasonable to do; hence, it's immoral.

When you universalize in this Kantian way, so too:

> **Lying is immoral because it contradicts truth-telling, and when you lie you depend upon the very idea of truth-telling so that people will believe your lies.**

> **Suicide is immoral because it contradicts self-preservation of one's life, and if everyone committed suicide there would be no life to kill, including your own.**

> **Giving discounts to friends at your store is immoral because it contradicts fair prices for all, effectively negating any possibility of giving discounts in the first place.**

It's a rationally-based moral theory through and through, and any kind of "performative contradiction"—like the ones mentioned above—is immoral. Kant is able to use this universalizability method to show that truth-telling, justice, and every other kind of right and good thing you can think of *really is* rational and, hence, moral; while lying, injustice, and every other kind of wrong and bad thing you can think of *really is* irrational and, hence, immoral.

> ## BOX 10-2 CRITIQUE IT!
> One problem with this universalizability method is that it is open to interpretation in a way that can be used by an agent to commit an immoral act, without engaging in a contradiction the way Kant claims. All one need to do is add the qualification "in this particular situation, right here and now" and the contradiction goes away. So, a thief could stop and think: What if everyone "in this particular situation, right here and now" were to steal this gold? It's not really *all of humanity* stealing at the same time, so to speak, so there is no inconsistency in the principle and, hence, it is not immoral.

We can paraphrase Kant's second formulation of his rational principle this way:

> Whenever you act, always treat yourself and others as an end in themselves, and never merely as a means to an end. In other words, don't ever use yourself or another person merely to achieve some other goal, no matter the consequences.

In other words, because they are conscious, rational beings, persons are precious in having an *intrinsic* value (as ends) and not an *instrumental* value (as a means to an end) like some object, tool, thing, or instrument. From this perspective, morally right decisions are those decisions where a person is treated as an end, and morally wrong decisions are those where someone is treated as a mere instrument or means to an end.

Why does Kant think people are intrinsically valuable? He argues that humans are unique in that they can recognize and act on reasons. Every other being (except perhaps God) is tied to instinct or natural law. Since humans can reason about their actions, they are actually a source of moral values. In virtue of being a source of moral value, they are themselves intrinsically valuable.

For example, take any "bad guy" you find on TV or in the movies. This bad guy almost always will use someone else to fulfill his nefarious plan to get money, revenge, or even take over the world. Recall, too, the cartoon with a villain (with cape, top hat, and curly mustache) who ties a girl to railroad tracks in the hope of destroying his nemesis, only to have his plan "foiled" by the hero who rescues her just in time, leaving the villain to cry "Curses! Foiled again!" The hero is like a good Kantian who tries to stop the bad guy or villain from using a person for his evil plan.

Of course, Kant doesn't say you can *never* use someone as a means—we need each other. He says never use someone merely as a means. What does it mean to treat someone also as an end? Some have interpreted Kant to mean that you should respect the intrinsic value of others by telling them the truth (do not deceive) and allowing them to freely make decisions (do not coerce). For instance, when you check out at the grocery store, you are using the cashier as a means to your end of getting food. But you are not using the cashier merely as a means; this person has agreed to work at the store for a certain amount of money. There is no coercion, because the cashier could quit if he wanted. There is no deceit, because the cashier was informed of the terms of the contract before agreeing to take the job. So, in paying for your food, you are treating the cashier as an end in himself, and not merely as a means.

Now, here's where Kant's claim that you should never use someone merely as a means is a little counterintuitive. Let's say a guy has a big bomb and will blow up a city of 250,000 people unless his old eighth grade teacher—whom the guy despises because he feels that the teacher is solely responsible for ruining his life—is brought out onto the steps of City Hall and executed. A Kantian would say that it would be immoral to execute the eighth grade teacher, even if doing so meant saving the city. This is because we would be using the teacher merely as a means to saving the lives of the city. The teacher could choose to lay down her life, but she is not obligated to do so, and no one can take it from her. In fact, even in this case, Kant says the moral blame for killing 250,000 people still falls on the guy with the bomb, not on the people being forced to make the decision. In forcing someone to make a decision like that, you have undermined his free choice, which renders him incapable of being responsible for his actions. Though this is perfectly consistent with Kant's theory, it is strikingly inconsistent with many people's intuition that we should kill one person to save 250,000.

Kant argues, further, that if a person's motivation or reason for a supposed moral act is anything other than dutifully acting from respect for morality—whether it be the good consequences that result from an action, benefits to self or others, your own desires or inclinations, or *even love itself*—then the act cannot rightly be considered moral. In fact, Kant says that Jesus' command in the Christian Scriptures to "love your enemies" is a clear case of dutifully acting in accordance with a moral principle and *not* acting according to your own inclination/desire, which is to hate your enemy. This does not mean that actions motivated by emotions *are not* morally good, only that we *could never tell* whether you are acting morally. If an act is consistent with duty, but motivated impurely, say, by emotions, Kant calls this an act "in accord with duty." If an act is consistent with duty and motivated solely by respect for duty, Kant calls this an act "from duty."

BOX 10-3 CRITIQUE IT!

Here is where Kant's moral theory has been critiqued for not being able to motivate people to be moral. How is a theory, that is, a purely rational explanation of the nature of moral action, going to motivate anyone to act? It seems you need to consider some consequences, some desire, or some kind of emotional *UMPH!* to get people motivated to act morally, as well as to prevent them from acting immorally. Often, it is precisely because we can imagine ourselves the victim of an immoral act—complete with all the negative feelings associated with it—that motivates us to do what is right and to not do what is wrong.

When we put these two formulations of Kant's moral principle together we get something like the following: Before we act, we need to check and see whether:

1. **Our act can be universalized.** Would we want our mom, our kid, or anyone to act this way, and, if so, is it consistent such that there is no performative contradiction?

2. **Our act respects the sacred dignity of a person** such that a person is never used merely as a means to another end.

For all its internal problems, Kant's theory was an advance in moral thinking, and still forms the basis of much of contemporary talk about moral principles and blind justice, as well as individual rights and respect for personal decision-making.

Consequence-Based Theories

In contrast to Kant's duty-based approach to morality, which doesn't consider the empirical consequences of an act, consequence-based theories argue that morality depends wholly on the consequences of an action. The most popular consequence-based theory is known as *Utilitarianism*, and it is the view that the morally right action is the act that brings about the most happiness for the most people over the longest period of time. One of the most prominent proponents of Utilitarianism was John Stuart Mill (1806–1873). Whereas a duty-based theorist would say, "It's the principle of the matter, and I am not concerned with the consequences to me or anyone else," a utilitarian would say something like, "I am concerned only with the consequences of the act to me and all affected, and I don't care about principles if they're going to bring about bad consequences."

Given the emphasis upon bringing about good consequences, utilitarians argue the *end* of bringing about good consequences justifies the *means* of doing something like lying or cheating in order to bring the good about. However, the view is not the "ends justify the means," because, stated this way, the "ends" could be anything, irrespective of others. Utilitarians are concerned only with one end, namely, more happiness for more people. "Happiness" usually translates as: more pleasure and less pain. This may sound very hedonistic, but it is not what we might call "classical hedonism." Classical hedonism might say that you should only do what brings you the most amount of immediate pleasure and the least amount of pain. This might entail that you should never do anything difficult or demanding, and that you should disregard others. Classical hedonism, as we're using the term, is also a consequence-based theory, known more widely as "Egoism."

Utilitarianism is different from classical hedonism in that it recognizes higher forms of pleasure, like art, music, and philosophy. Working hard to learn a language may yield a qualitatively greater pleasure than getting drunk or smoking pot. Similarly, helping others is more satisfying than only looking out for yourself. In fact, Mill argues that anyone who has experienced both higher and lower forms of pleasure shows a marked preference for the higher pleasures. And people who help others are generally happier than people who don't. Therefore, Utilitarianism, he argues, will not lead to a purely egotistical and carnal society.

In opposition to the duty-based view that persons never should be used merely as means to some end, consequence-based theories sometimes justify treating others as means to the greater good of happiness for a majority of people. If killing one, two, or even a hundred people is required to save a significantly larger group of people, then, on utilitarian grounds, this killing is morally obligatory. However, most utilitarians argue that people are rarely faced with such unfortunate dilemmas, and that acting for the most pleasure for the most people actually results in many of the same actions as traditional duty-based theories, though for different reasons.

For instance, a duty-based theorist has an obligation not to lie since lying would use an intrinsically valuable person as if they were merely extrinsically valuable—that is, valuable for the liar's end. But a utilitarian may also have a moral obligation not to lie, since lying often leads to more interpersonal conflict than it resolves. Therefore, telling the truth may, more often than not, lead to the most happiness for the most number of people involved.

The way in which the utilitarian determines the good consequences to all affected in a situation is through a pro vs. con kind of calculus, or adding up all of the gains/pleasures on one side and comparing them with all of the losses/pains on the other side. The moral decision, then, is the one where the most benefits/pleasures will result. Also, since everyone is considered completely equal in terms of their worth, it appears to be a fair way to determine actions. We can think of the decision-making process of a utilitarian like a cost-benefit analysis used in many business decisions (see Fig. 10-5).

INTENDED ACT	GAINS	LOSSES	WHAT SHOULD BE DONE
Don't do homework and lie about it to parents	• gets me out of homework • can watch TV • short-term pleasure	• get in trouble with parents • get in trouble with teacher • get behind on lesson • longer-term pain of not knowing material	Actually do homework
(Kayaker dies in rapids.) Bring up his dead body that is stuck below water in rapids	• closure for family	• spend lots of time and money • likely kill professional diver(s)	Leave body in water
Shoot man who is likely to kill family he has hostage	• save lives of rest of family • ends stand-off, which is scary, stressful for all • ends all time and money spent on situation	• kills man • leaves family without father	Shoot the man

FIGURE 10-5 · Interpreting the utilitarian calculus as a cost/benefit analysis.

So too:

> If a company is going to lose lots of money and all people will lose their jobs (which will cause a lot of pain to a lot of people), then the moral decision might be to lay off a few workers (which will cause pain, but not as much) for the overall benefit of the company.

> If a runaway train is heading down a track where there are ten workers who will be killed (a big loss), and the engineer can steer the train onto another track where there is just one worker who will be killed (a loss, but not as big), then the engineer is morally obligated to steer the train so that one person is killed.

> It may be necessary to shoot down a hijacked plane to prevent it from crashing into a building where there are lots of people.

> If a few people are offended by a work of art on display, but many more enjoy it, then the morally right decision is to continue displaying the art.

BOX 10-3 CRITIQUE IT!

Utilitarians are often criticized because it just seems that the weighing pros and cons approach is too simplistic and counterintuitive. It's hard to say exactly: (a) what the "weight" of one pleasure/pain is, versus another pleasure/pain, as well as (b) what the future will hold in terms of consequences (they might turn out differently from what you think). Also, consider these dilemmas:

1. If you know that by dating woman A, women X, Y, and Z will commit suicide, should you not date woman A? What would the utilitarian have to do?

2. If most people in a group are racists, and it brings them pleasure to be so, should racism be proffered and encouraged? Again, what would be consistent with utilitarianism?

3. If we can introduce a drug into a society that will totally placate people such that they are completely moral, alleviating all strife, conflict, pain, and evil, should we introduce that drug?

Character-Based Theories

Opposed to both duty-based and consequence-based moral theories—which try to establish what people should *do* and then assess whether they have actually done so—there are *character-based theories*, which evaluate the value of a person's *character* and then assess which actions best contribute to a certain kind of character. The most popular character-based theory is Aristotle's *Virtue Theory*. The central idea of virtue ethics is that character and behavior are mutually forming. Having a certain type of character will determine how you act, and how you act over time will determine what kind of character you have. If someone has a virtuous character, then they will often do what is best (though what is "best" is not clearly defined in virtue theory—they do not talk about duties or consequences). Similarly, if someone does what is best over time, her character will develop in such a way that she will be more likely to do what's best in the future. And, after all, we want to not only perform right actions (duty-based) that have good consequences (consequence-based), we also want to be *virtuous people* performing right actions that have good consequences.

> ### BOX 10-4 Character Counts
> Aristotle implies that, though you may be able to convince a demon to do the right thing (according to some duty) or to bring about good consequences (according to the utilitarian), the agent is still a demon. Thus, virtue ethics can act as a kind of complement to the Kantian and utilitarian positions, rounding out our moral lives.

The idea here—which was noted by philosophers as far back as Confucius (551–479 BCE) and Plato (427–347 BCE)—is that different people have characters which are formed through their interaction with their environments. Aristotle (384–322 BCE), in particular, thought that character development was the key to morality. As he saw it, our characters result from (1) forming certain good habits starting in childhood and (2) acquiring practical wisdom in maturity. *Virtue* is a good habit whereby one fosters a kind of balance in one's character. The idea is to promote "not too much" and "not too little" of some character trait, but "just the right amount," so that our actions and reactions reflect appropriate character. *Virtues* are the traits that fall within the mean, or average, of the extremes of too much (the vice of excess) and too little (the vice of deficiency). Virtue ethicists identify a general list of virtues, including honesty, courage, prudence, generosity, integrity, affability, and respect, to name just a few.

Here are some standard examples of virtues, with their two vices (Fig. 10-6):

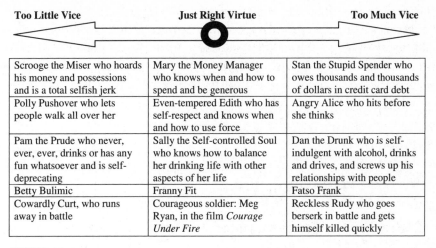

Too Little Vice	Just Right Virtue	Too Much Vice
Scrooge the Miser who hoards his money and possessions and is a total selfish jerk	Mary the Money Manager who knows when and how to spend and be generous	Stan the Stupid Spender who owes thousands and thousands of dollars in credit card debt
Polly Pushover who lets people walk all over her	Even-tempered Edith who has self-respect and knows when and how to use force	Angry Alice who hits before she thinks
Pam the Prude who never, ever, ever, drinks or has any fun whatsoever and is self-deprecating	Sally the Self-controlled Soul who knows how to balance her drinking life with other aspects of her life	Dan the Drunk who is self-indulgent with alcohol, drinks and drives, and screws up his relationships with people
Betty Bulimic	Franny Fit	Fatso Frank
Cowardly Curt, who runs away in battle	Courageous soldier: Meg Ryan, in the film *Courage Under Fire*	Reckless Rudy who goes berserk in battle and gets himself killed quickly

FIGURE 10-6 · Virtues and vices.

The virtuous person has cultivated the kind of character whereby she knows how to act and react in the right way, at the right time, in the right manner, and for the right reasons in each and every moral dilemma encountered (in some vague sense of "right").[†] However, the way in which one cultivates a virtuous character is through choosing actions that are conducive to building that virtuous character. So for example, if one wants to cultivate the virtue of honesty so that one can actually be an honest person, then one needs to *act* honestly time and time again so that the virtue can "sink in" to the person's character. The more Johnny actually tells the truth when asked whether he has done something wrong, the more Johnny cultivates the virtue of honesty. The more Suzy lies when asked whether she has done something wrong, the more she cultivates the vice of dishonesty.

> **BOX 10-5** Virtue Ethics vs. Utilitarianism
>
> A virtue ethicist could argue that the kind of "using of a person for the greater good" aspect of utilitarianism is immoral. Respect is the key virtue here, and several virtue ethicists have argued that objectification stems from a disordered or unbalanced character. The person who has cultivated respect for persons in his

---

[†]To be fair to Aristotle, he defines what is "right" as "achieving the goal appropriate for you," that is, your "end." And the end for humans is to act virtuously. There are a number of questions that can be raised about this view, including: Why is virtue "good"? How do you know what ends a human has? How do you know that humans all have the same ends? Aristotle attempts to answer these questions in his monumental work, Nicomachean *Ethics*.

or her character naturally will not objectify another person. When one treats a person as an object, one empties another of their intrinsic dignity, value, and worth, affecting both the one doing the objectifying and the one being objectified. In effect, the problem lies in the psychological ill-effects of treating another as less than a person.

Still Struggling

Character vs. Consequences vs. Duty

Using a person or sacrificing oneself to save a group of people is one thing; using a person for your own personal pleasure or gain is another. We honestly must ask ourselves these four questions:

1. As a fully rational and autonomous person, may I treat myself or another fully rational and autonomous person only as a means, rather than as end in him or herself?

2. What kinds of consequences will result for other persons affected by my action and me if I do decide to treat myself or another fully rational and autonomous person as a means, rather than as an end in him or herself?

3. Do I want to foster a virtue in myself, my kids, my family, my community, and/or in my world whereby others are seen as persons worthy of respect, fundamentally equal to myself, making it such that one person is not permitted to objectify another person?

4. Do I want to foster a vice in myself, my kids, my family, my community, and/or in my world whereby certain others are seen as emptied of the intrinsic dignity, value, and worth, such that I am permitted to objectify another person?

Of course, these questions are "loaded" to a great extent; but this is to make a point. It seems that, if (1) and (2) do not halt us in our tracks and stop us from using ourselves or one another, then (3) and (4) give us pause to consider what kind of person we would becoming if we continually use others.

Character-based theories face problems, however, just like duty- and consequence-based theories. First, Virtue Ethics is not, by itself, a decision procedure. So, it cannot tell you whether an act is permissible, impermissible, obligatory, or supererogatory, only whether it is virtuous or vicious. Because of this, there is no moral praise or blame. Therefore, in order to be plausible, Virtue Ethics must be combined with a theory that includes a decision procedure.

Second, Virtue Ethics gives no account of why the virtues are virtuous. Aristotle explains that being virtuous just means achieving your predetermined goals. He then says that your predetermined goals include being honest, just, friendly, cautious, wise, etc. But there seems to be a gap between his definition of virtuous and the virtues themselves. Why are *those* the virtues, and not some other set of behaviors? It would seem the virtue theorist cannot answer this question.

QUIZ

1. Historically, there has been no difference between ethics and morality.
    A. True
    B. False

2. Abortion is an issue found primarily in
    A. normative ethics.
    B. metaethics.
    C. cyberethics.
    D. none of the above.

3. _____ act is one in which no positive or negative moral implications ensue.
    A. A negligible
    B. An obligatory
    C. A permissible
    D. A good

4. _____ undermines meaningful discourse. Without meaningful discourse, any investigation into reality is doomed.
    A. Vagueness
    B. Falsity
    C. Inconsistency
    D. Incongruence

5. Kant says that you can *never* use someone as a means.
    A. True
    B. False

6. The way in which the utilitarian determines the good consequences to all affected in a situation is through
    A. research.
    B. checking with experts.
    C. both of the above.
    D. a pro vs. con kind of calculus.

7. If we can introduce a drug into a society that will totally placate people such that they are completely moral, alleviating all strife, conflict, pain, and evil, should we introduce that drug? This question arose in the context of
    A. duty-based moral theory.
    B. virtue-based moral theory.
    C. consequence-based moral theory.
    D. none of the above.

8. **Making your little brother do his homework arose in the context of**
   A. duty-based moral theory.
   B. virtue-based moral theory.
   C. consequence-based moral theory.
   D. none of the above.

9. **Acting from respect for morality arose in the context of**
   A. duty-based moral theory.
   B. virtue-based moral theory.
   C. consequence-based moral theory.
   D. none of the above.

10. **Virtue Ethics is, by itself, a decision procedure.**
    A. True
    B. False

chapter **11**

Moral Decision-Making

CHAPTER OBJECTIVES

In this chapter, you'll learn about…

- How to evaluate moral claims
- Identifying relevant moral principles
- Identifying relevant consequences of moral actions
- Applying moral theories
- Applying a reasoning strategy to moral actions
- Moral relativism
- Religious-based morality

In the previous chapter we explained that moral philosophy, or ethics, is the branch of philosophy that investigates and critiques:

(a) **human actions that affect morally relevant beings (definitely humans; probably many animals) and**

(b) **the principles that people appeal to when they act.**

We also explained that moral philosophers have put forward principles that people should follow when making a decision, and that these principles are justified by moral theories. But once we have a plausible moral theory, how do we move from this theory to a decision? How does a moral theory help us know which acts are right and which are wrong?

How to Evaluate Moral Claims

To answer this, it will be helpful to walk through a typical moral decision-making process. There are a variety of different strategies for making morally-informed decisions, and we will give you an example of our favorite. It is by no means exhaustive and, although there is a sequence to the steps, you may need to move back and forth among them as you obtain new information. For clarity, we'll walk you through the process by applying it to a controversial moral decision. Take abortion, for example. Is it morally permissible to abort a fetus?

1. Clarify Your Terms

As with almost any kind of decision, it's important to clarify any terms you will be using.

- Is a "human embryo" a blob of cells? Yes.
- Is the blob of cells we're talking about *human*? Yes, in the biological sense, it is human; it has human DNA.
- Does "biological humanity" matter for the sake of morality? My fingernail also has human DNA, but it is not immoral to dispose of fingernails. So no, DNA is not the issue.
- Is the blob of cells *only* a blob of cells? That's what we're interested in, so we can't say just yet.

There are many other terms to clarify, including, "abortion," "conception," and "person." And one can imagine what kinds of tragedy can occur in our moral decision-making if we *don't* define our terms correctly. Someone who defends a "pro-choice" political position might consider a human embryo merely a blob of cells and feel justified morally in aborting, while someone who defends a "pro-life" political position might consider a human embryo something that should be protected and feel that few abortions are morally justified. On one hand, if the pro-choice defender is right, then a woman's liberties might be infringed upon by a pro-choice policy. On the other hand, if the pro-choice defender is wrong, then many innocent people may be murdered. The issues involved in the abortion debate are serious, and terminology is incredibly important. With a set of clear terminology about pregnancy and abortion, the later steps in the decision-making process are much easier.

Here are some further examples of issues where identifying clear definitions affects your moral decision-making:

- One definition of a *just war* is: a war whereby one country is justified in defending itself against another on grounds of self-defense. As a budding philosopher, you need to evaluate as to whether this is even a good definition of a just war. Further, why does "self-defense" justify violence? And, does a country's right to defend itself include killing?

- A pacifist might maintain that any instance of one human killing another is murder and murder is wrong; therefore, no humans should ever kill other humans. But, what is *murder*? Is it the *unjust* taking of another human life with malicious intent? If that's what it is, then is killing in self-defense really murder? Are all "pacifists" committed to the claim that no humans should ever kill other humans?

- Ex-U.S.-President Bill Clinton maintained that he "did not have sexual relations" with former White House intern Monica Lewinski. What did he mean by "sexual relations"? Apparently, for Clinton, it includes actual intercourse but not oral sex. He eventually admitted to having oral sex with Lewinski.

2. Get the Facts

As with any decision, it is important to first get all of the facts. This is not always easy to do, and in many cases we cannot obtain all the relevant information. Let's say that Johnny kicked a ball and hit Sally in the face. Is Johnny morally blameworthy for doing so? That depends on a number of non-moral facts about the situation. Did Johnny *intentionally* kick the ball in Sally's face or was he just kicking it around the schoolyard without looking where he was kicking it? If he did intentionally kick it, was Sally antagonizing him (which may not fully justify him, but it may mitigate punishment). If he did not intentionally kick it, was he apologetic (a lack of proper affect may be morally blameworthy)?

Returning to our abortion example, before making a moral judgment on abortion, it is important to acquire the available facts about the development of a sperm and egg from a zygote into a blastocyst, which then develops into an embryo, which then develops into a fetus. When do the sperm and egg become an individual human? What if you learned that the genetic material of the two doesn't fully merge until after 24 hours? And what if you learned that there is not a unique set of human DNA until sometime between 8 and 12 days? Would these facts affect your belief about the moral permissibility of the "morning after" pill? Similarly, we know that a fetus begins feeling pain at around 20 weeks. If pain is morally significant, would this change your belief about when, if ever, some types of abortion are permissible?

3. Identify Relevant Moral Principles

After clarifying your terms and identifying the relevant non-moral facts, you can begin assessing the moral implications of your possible decisions. Once the situation is clear, certain beliefs will emerge that have a compelling sense of duty associated with them. These are your initial moral intuitions about the case: "I shouldn't do that" or "It is permissible for me to do that." At this stage it is important to ask *why* you think this way. Upon reflection, you can identify a set of principles that justify these beliefs, for example, "People should not harm others without good reason," and, "It is wrong to break something that doesn't belong to you."

For example, if you are in a grocery store and see someone stuff a candy bar in his jacket and walk out without paying for it, you may intuitively think, "That's not right." If you ask why you think this way, you could respond, "People should not take things that do not belong to them; they are not entitled to it." The more general principle explains why you believe the thief has done something wrong, and, if the principle is true, justifies your belief that the thief is wrong. That is, if the principle is true, the thief has really done something immoral, and may be blameworthy for doing so. (To be sure, he may be guilty without being blameworthy, for instance, if he were mentally handicapped and, therefore, could not consider reasons for acting or not acting.)

The grocery theft example highlights one example of a moral principle, but there are many others like it. The following are moral principles that hold virtually worldwide:

- One should not cause unnecessary harm. (One should prevent unnecessary harm, if possible.)
- One should not kill without sufficient reason. (One should not murder.)
- One should defend oneself against an attacker.
- One should not deceive. (One should tell the truth.)
- One should not steal.
- One should prevent others from stealing.
- One should not rape.
- One should care for children.
- One should play fairly.

These principles are widely held to be true. But even so, do they hold in every conceivable circumstance? Clearly not. In fact, sometimes they conflict. A classic example asks you to consider living in Nazi Germany during WWII. Imagine you are hiding Jewish people in your house and the Gestapo comes to the door and asks whether you are hiding Jews. It would seem that two principles are in conflict: one should prevent unnecessary harm and one should not deceive. But it seems clear that lying is a morally good thing to do here; in fact, you may be morally obligated to lie in this situation.

Consider, again, our abortion case. What principles are relevant? It would seem that the principles "do not harm unnecessarily" and "do not murder" are certainly relevant. These may conflict with a principle about a person's right to behave how she wishes, for instance, "do not arbitrarily restrict someone's freedom." If abortion is murder, the two principles do not actually conflict, since prohibiting someone from committing murder is not an *arbitrary* restriction. If abortion is not murder, the two principles also do not come into conflict, since killing a blob of cells is not an instance of murder. The difficult part of many moral decisions is determining which principles apply and how. For instance, a fetus will be harmed by abortion after 20 weeks. Does this harm outweigh a person's right to behave how she wishes, or do her wishes outweigh the harm?

As you can see, we will almost inevitably discover conflicting moral principles. How do we know which of two conflicting principles to rely on? What reasons do we have for thinking that any of these principles is true, or at least more relevant for any particular decision? To see how this, works, move to the next step in our decision strategy.

4. Apply Moral Theories

Consider a doctor doing a routine hernia operation on a 70-year-old female patient. Let's say the doctor discovers that she has an undescended testicle. The woman has never been able to have biological children of her own, and the doctor knows that this has been a real source of psychological trauma for her. In fact, the doctor truly believes that, given her age and psychological state, this news about the undescended testicle might throw her into a depression from which she would not recover. Should he tell her and risk violating the principle "Do no harm"—since she'll be devastated—or should he say nothing and (arguably) violate a principle like "Doctors ought to reveal facts that they discover about their physiologies to patients?"

Which principle takes precedence? To answer, we need a clear understanding of the most plausible moral theories. In the previous chapter, we discussed three: duty-based theories, consequence-based theories, and character-based theories. Recall that, according to duty-based theories, a person's inherent value as a moral reasoner establishes certain things you cannot do to her and certain obligations you have toward her—for instance, truth-telling and promise-keeping. In the case of the doctor, his duty to mitigate harm is overridden by his duty to respect the patient as a competent moral reasoner. Even if telling her about the undescended testicle is emotionally painful for her, it is better that she knows it and can deal rationally with it, rather than not knowing. And, who knows? Knowing why she couldn't have children may be more comforting to her than remaining in doubt. Since the doctor can't be sure how she will react, but can be clear that not telling her is a form of deceit, he should take the risk of telling her.

In the abortion case, the question of whether or when the fetus can feel pain or be harmed by the abortion is not relevant to the duty-based theorist. Moral agents must take responsibility for their actions, and if someone acts in such a way that will produce a being that will eventually be able to consider and act on moral reasons, she should not treat that being merely as a means to her own personal happiness. Kant argued that, even though animals are not moral agents, it is not consistent with our unique moral status to treat them cruelly, since doing so serves no rational purpose. Some contemporary Kantians have also applied this reasoning to fetuses. We should not dispose of them flippantly any more than we should dispose of animals flippantly.

Consider, instead, the consequence-based theorist, who argues that a being's capacity for suffering establishes how you should and should not act toward it. In the doctor's case, preventing harm to the patient takes precedence over her knowledge of her own condition. Her undescended testicle is irrelevant to her hernia operation, and it is highly likely that telling her will lead to much pain. Not telling her will result in no extra pain. Therefore, it is morally better for the doctor to remain silent. The likelihood that she will be more comfortable as a result of knowing why she couldn't have children is lower than that she will be depressed. With respect to abortion, whether the fetus can experience pain is the primary, if not the only, consideration. After 20 weeks, fetuses begin to exhibit pain-avoidance behavior, just like you and us. We can't be sure they're in pain, but we know the neurophysiology is intact for pain sensation and we know they act as if they were in pain. Therefore, abortion after 20 weeks is probably impermissible. On the other hand, if the pain of having the child (financial hardships, etc.) is greater than the pain the fetus will suffer, abortion may be obligatory even after 20 weeks.

How might character-based theories weigh in on these decisions? Remember, character-based theories do not offer decision-procedures. They might, however, be useful for choosing among moral theories that do include decision-procedures. For instance, which of the decisions suggested by the two types of theory above are more likely to lead you to act according to that theory consistently in the future? Would the doctor's deceit lead him to be less honest overall with his patients? If so, he should probably tell her about her condition. If not, perhaps it is permissible for him to refrain from telling her. Would the doctor's telling her lead him to be callous and cold with future patients? If so, he should probably refrain from telling her. If not, perhaps it is permissible for him to tell her. Character-based theories seem to suggest that the best course of action for you depends on your psychology and which reasons for action will foster good actions in the future.

Apply each theory to see how each would lead you to act. This will also help you recognize alternative actions you might not otherwise think of. After you have considered each set of options, determine which moral theory is most plausible. How do you decide which is most plausible? This is the hard work of moral philosophy, and there is widespread disagreement. However, the tools of the best moral philosophers are available to you in the next step.

Still Struggling

Applying a moral theory to a decision can be tricky, especially given the vast number of relevant considerations. Here are a couple of tips that should help keep your decisions well within the range of the most plausible moral theories. Ask first: If I perform act X, will anyone's rights be violated? "Rights" are a tricky subject in themselves (see chapter 12), but think of Kant's conception of using someone merely as a means as a restriction on rights. Many contemporary utilitarians would agree with contemporary Kantians that protecting these rights leads to more happiness most of the time. If rights are violated, there are good (though maybe not conclusive) reasons for not performing act X. If no rights are violated, ask next: If I perform act X, will anyone be harmed? If the answer is yes, then you must consider whether the good to be gained outweighs the harm caused. If no rights are violated and many are benefitted, some harm may be justified. If no more benefits are likely than not performing the act, then the fact that it causes harm is a reason to refrain from performing it.

These guidelines attempt to employ the best of both the Kantian and utilitarian moral theories. But we must be clear: **This does not mean they can both be true.** They are mutually exclusive accounts of what makes an act permissible or impermissible. If you are a Kantian, considerations about harm will take the form of duties against unnecessary harm. If you are a utilitarian, considerations about individual rights will be considered within the broader questions of overall happiness.

5. Apply a Reasoning Strategy

As rational beings, it's probably understood that people should think critically, clearly, and consistently about any decision they make. Moral theories are sets of internally consistent criteria about what features of reality make an act right or wrong. Because they include these criteria, moral theories can guide our decisions; that is, each constitutes a comprehensive decision-procedure for moral action. But not all theories are good theories. Even if they are internally consistent, they may be inconsistent with the best evidence we have about what is right and wrong. This problem is not particular to ethics. Science has faced this problem for centuries.

Copernicus' physical explanations were better than Ptolemy's, and Newton's were even better than Copernicus'. But even Newtonian physics was inadequate for some questions, and Einstein showed why—Newton had an inaccurate understanding of relativity and used the wrong geometric system. Was Einstein right? About some things, it would seem, but not others.

So, how do we deal with inadequate theories? We choose the most plausible version we have and try to fix it. How do we know whether our theory is adequate? We test it, using argument strategies. Recall from Chapter 2, we discussed direct arguments, thought experiments, inference to the best explanation, and fallacies. If a theory commits a fallacy or falls prey to a counterexample, it faces a serious problem. We should either attempt to revise the theory to avoid this implication or we should reject the theory in favor of one that doesn't face this problem.

What does it look like to test a moral theory? Classical versions of duty-based theories and consequence-based theories face interesting problems. Here is an example of each:

Some duty-based theories include the claim that our duties toward an action (say, "Do not lie.") are absolute, that is, there is no instance in which it should be violated, even if it leads to worse consequences. But there seem clear cases where our duties come into conflict. Here is a counterexample (premises 1–3 can be found in Immanuel Kant's *Groundwork for the Metaphysics of Morals*):

1. Moral duties are absolute.
2. No one should break a promise to prevent harm.
3. Everyone should help another when she can.

Now, imagine I am faced with the following decision. I have promised to meet my mother for dinner at 7. On my way to dinner, around 6:45, I see a child get hit by a car that leaves the scene of the accident. No one is around, and if the child is to receive medical help, it is up to me. But if I stay and help, I will miss my dinner date with my mother. So, I have a new premise:

4. I should either stay and help, violating my duty to not break a promise to prevent harm, or I keep my promise, violating my duty to help another when I can.

Now I am faced with a situation in which I am forced to violate an absolute moral duty. Something has gone wrong. An ethical theory that entails conflicts among obligations is not a very good decision-procedure. So, how might we fix it? It seems right that, in most cases, it is bad to lie, and that, in most cases, we should help people in need when we can. Therefore, the most plausible premise to reject is (1). However, this has serious implications for the rational justification for the theory, so we must be careful not to undermine the whole system in rejecting a major tenet. Many contemporary moral philosophers offer suggestions for how to fix duty-based theories, and they are worth attention.

A similar problem faces consequence-based theories. Utilitarian consequentialism includes the claim:

1. An act is right insofar as it increases overall happiness for the most people and wrong if it decreases overall happiness.

But consider the following decision. Imagine you are flying a plane that has engine trouble and you are being forced to make an emergency landing. Your only options for landing the plane in reasonable condition are the heavily traveled interstate or a large grassy field. On one hand, it is rush hour, so there are too many cars on the interstate to prevent crashes and several casualties. On the other hand, right in the middle of the field where you would need to land your children are playing with the family dog. You face the following decision:

2. I should either land on the interstate causing a great deal of harm to a large number of motorists or land on the field killing my children, causing much less harm overall.

3. Landing on the field causes less harm, so, according to 1, I should land on the field.

There is no contradiction here, but it is inconsistent with our moral intuition that we have duties to our children that override good consequences for others. Interestingly, this view also entails that if a majority of people would be much happier than a minority if the majority forced the minority into working for them for free, slavery would be morally obligatory. How might we fix this theory?

Something that would help would be to remove the emphasis on increasing happiness and focus on not decreasing it. If my only obligation is not to cause you harm, then I am not obligated to hurt others to accomplish that. I can just avoid you, or take extra care when I interact with you. Unfortunately, this won't help the airplane dilemma. Even if I focus on not decreasing happiness, I am still obligated to kill my children rather than the motorists. Like duty-based theories, many contemporary moral philosophers offer suggestions for fixing consequence-based theories, and they too are worth attention.

6. Make A Decision

After doing all the hard work of reasoning about the morality of a decision, follow through with your decision. You have clarified your terms, obtained the relevant facts, identified some relevant moral principles, and applied the best available moral theory. You have almost all the information you could possibly have about that decision (unless you missed a few facts). Therefore, no one is in any better position than you to decide what to do in this situation. This means that it would be irrational to act in a way that is inconsistent with the conclusion of your decision procedure. Committing to a decision is often one of the most difficult aspects of moral reasoning, but it is no less important than any other. Especially since *not doing anything* is often a decision. Ignoring the problem is rarely an option. Below you can see a summary of our decision procedure (Fig. 11-1).

When faced with a moral decision…

1. Clarify Your Terms
2. Get the Facts
3. Identify the Relevant Moral Principles
4. Apply Moral Theories
5. Apply a Reasoning Strategy
6. Make a Decision

FIGURE 11-1 • A decision procedure for answering moral questions.

Two Problematic Moral Theories

Someone might rightfully ask, "Why should I restrict my decision-making process to the three moral theories you mentioned? Aren't moral claims relative?" or "Doesn't God tell us what is right and wrong?" These are legitimate questions and deserve a response. Few moral philosophers hold these positions because they face problems that are more serious than the three traditional moral theories we discussed in the last chapter. To show why these problems are more serious, we'll introduce a new reasoning strategy called *reductio ad absurdum* (Latin for "reduction to absurdity"). In a *reductio ad absurdum*, the claims of the view are included in the premise, and then a contradiction is derived from them, like this:

1. If this position were true, then a contradictory conclusion results (an absurdity).

2. But, we cannot accept contradictions because no contradictions are true.

3. Therefore, the position can't be true, that is, it's false (and one should not hold such a position).

We'll briefly explain why the suggestions that moral claims are relative or depend on God are not as plausible as the three moral theories we have been focusing on.

Moral Relativism

We all agree that there's a variety of different cultures on this planet with people who believe a lot of different things about themselves, their communities, and reality as they perceive it. People from different cultures also believe different (sometimes contradictory) things about what is moral and immoral. This is not a problem in itself; it just means that they do not all hold correct moral claims. The problem is that it would seem that no one is in any position to say just what "correct" moral claims are. We are all tied to our culture in important ways, and our moral views are shaped partly by our cultures. So, why not think morality is just a matter of culture?

Here's a version of an argument for moral relativism:

1. Each culture has a set of values that it deems important and generates moral laws as a result of those values.
2. When you investigate the cultures of the world, you discover that cultures have different and even contradictory sets of values and moral laws.
3. No one can consider the truth of a moral claim outside the context of a culture.
4. Thus, the truth of all moral claims is relative to the culture in which it is considered.

This sounds pretty powerful on the face of it. But it entails a contradiction. Consider the following argument:

5. Moral relativism is true: The truth of all moral claims is relative to the culture in which it is considered.
6. Some cultures believe that the truth of moral claims is universal (it holds in all cultures).
7. Therefore, in that culture, moral relativism is false.
8. If moral relativism is not true of all cultures, then the truth of every moral claim is not relative to the culture in which it is considered.
9. Therefore, cultural relativism is false.

If moral relativism is true, it entails its own falsity. Therefore, it is self-defeating, and we should not accept it.

BOX 11-1 CRITIQUE IT!

Moral relativism suffers from a number of other problems, as well. Here are two more examples:

1. First, premise 2 above has been challenged by empirical evidence supporting what is known as *soft universalism*. Soft universalism is the belief that there is a small set of core values that hold no matter what culture you're in. Today, many cultural anthropologists who have actually lived in numerous cultures defend a version of soft universalism. This small set of core values gives rise to a small set of moral laws that then can be said to hold universally in the world, no matter what the culture. This set of moral claims includes:

   a. Don't murder

   b. Don't lie

   c. Don't harm needlessly

   d. Protect your children

2. Second, moral relativism faces the practical objection that, if it is true, a person from one culture cannot say to a person from another culture that what she is doing is right or wrong. Again, if there aren't any moral laws that hold universally, then no universal judgments can be made. If this is true, then no one can object on moral grounds to genocide, mass rape, unjust displacement of people, torturing and killing of innocents, stoning of women who seek justice, and every other kind of horrible harm we can think of that takes place in another culture. Therefore, no one can morally judge Hitler or Stalin for killing people for personal gain. Their culture approved their actions. Similarly, if your own culture approves of rape or genocide, you have no moral grounds on which to critique it—in objecting, according to relativism, you are suggesting that people act immorally. From the practical perspective of forming and developing moral beliefs, this seems unacceptable.

Religious-Based Morality

Many people on the planet form their beliefs about morality and moral value through the framework of their religious traditions. Because of this, some have claimed that morality must be something that God decides. God (or some relevantly-similar divine being) is all-good, all-knowing, and all-powerful, and without him, there is nothing; therefore, if any moral claims are true, God must make them true. He created humans, and then set the rules for human behavior. This is a view known as "Divine Command Theory."

A problem for Divine Command Theory is that it also seems self-defeating. Consider the following argument (motivated roughly by Plato's dialogue, *Euthyphro*):

1. God either makes moral claims true by declaring them true, or declares they are true because they are already true.

2. If God makes them true by declaring them true, then he is not all "good," since the term would not mean anything until he said it, and would then be a matter of preference (if God had said "Murder is obligatory," then it would be).

3. If God declares they are true because they are already true, then he is not all-powerful, since there are some true claims that restrict what God can do (God can't claim the right to kill just anyone he wants too).

4. Therefore, either God is not all-good or not all-powerful, or there are no true moral claims.

This conclusion seems to leave people who believe in God and moral truth in a bad spot. Thankfully, there have been a number of very good responses to this dilemma. They do not, however, typically end with the traditional Divine Command Theory. Even C. S. Lewis, one of the twentieth century's premier representatives of Christianity, wrote:

> It has sometimes been asked whether God commands certain things because they are right, or whether certain things are right because God commands them. …I emphatically embrace the first alternative. The second might lead to the abominable conclusion…that charity is good only because God arbitrarily commanded it—that He might equally well have commanded us to hate Him and one another and that hatred would then have been right (*The Problem of Pain*, 1940:99).

Lewis believes that God's will is constrained, but not in a way that undermines his power. "God's will is determined by His wisdom which always embraces, the intrinsically good" (1940: 99). This means that morality is like mathematics. God can't make $2 + 2 = 5$, but this is no smear on his power; it is actually a virtue that he can't do absurd things. In the same way, God cannot violate absolute moral principles because he perceives them as perfectly good. And this is what makes him good. So, it would seem that the existence of both God and moral truth are compatible, but this doesn't help Divine Command Theory; this view seems untenable.

There is also one additional reason to resist religious-based moral theories. Even if Divine Command Theory were correct, how might we fumbling humans figure out which God commanded which moral truths? We cannot go to our religious texts. No sacred texts mention identity theft, intellectual property, or emotional affairs. In addition, religious texts disagree on what acts are moral just like different cultures do. So, where do we go for answers? It seems we are left with the traditional moral philosophers, attempting to *reason* about which acts are right and wrong. And it seems the best candidates for such a theory are the three we discussed in the preceding chapter.

QUIZ

1. "Get the Facts" is the first step in the moral decision-making process.
   A. True
   B. False

2. Returning to our abortion example, before making a moral judgment on abortion, it is important to acquire the available facts about the development of a sperm and egg from a _____, which then develops into an embryo, which then develops into a fetus.
   A. cyst into a zygote
   B. blastozygote into a blastocyst
   C. zygote into a blastocyst
   D. blastocyst into a zygote

3. In this step of the moral decision-making process, certain beliefs will emerge that have a compelling sense of duty associated with them.
   A. Get the Facts
   B. Clarify Your Terms
   C. Identify Relevant Moral Principles
   D. None of the above

4. In the case of the woman with the undescended testicle, if the doctor lied to her, he might be considered to be upholding the following principle:
   A. "Do no harm."
   B. "Doctors ought to reveal facts that they discover about their physiologies to patients."
   C. "Doctors ought to reveal facts, only if those facts are worth revealing."
   D. None of the above.

5. _____ argued that, even though animals are not moral agents, it is not consistent with our unique moral status to treat them cruelly, since doing so serves no rational purpose.
   A. Mill
   B. Aristotle
   C. Hume
   D. Kant

6. "Get the Facts" is the last step in the moral decision-making process.
   A. True
   B. False

7. _____ do not offer decision-procedures. They might, however, be useful for choosing among moral theories that do include decision-procedures.
   A. Character-based theories
   B. Consequence-based theories
   C. Moral theories
   D. Duty-based theories

8. _____ argues that a being's capacity for suffering establishes how you should and should not act toward it.
   A. A character-based theorist
   B. A consequence-based theorist
   C. An amoral theorist
   D. A duty-based theorist

9. Not all theories are good theories. Even if they are internally _____ , they may be _____ with the best evidence we have about what is right and wrong.
   A. inconsistent, consistent
   B. valid, invalid
   C. consistent, inconsistent
   D. stable, unstable

10. Moral relativism faces the practical objection that, if it is true, a person from one culture cannot say to a person from another culture that what they're doing is right or wrong.
   A. True
   B. False

chapter 12

Justice, Rights, and Government

CHAPTER OBJECTIVES

In this chapter, you'll learn about...

- Different kinds of government
- Paternal and liberal governments
- Political liberty
- Laws and morality
- Individual rights vs. the common good
- Justice

What kind of government should a society have? Should it have a lot of control over its citizens, like a parent guiding her children? Or should it have very limited control, and only step in to the affairs of citizens when absolutely necessary— for example, to adjudicate conflict, prevent harm, or address grievances? What exactly is the nature of justice, laws, and rights? In a society, should you have the right to express yourself freely in any way you want to, or should your freedoms be restricted by the needs of others? Should the rights of one person override the common good of all, or vice-versa? Should there be a welfare system? Capital punishment? Gun control?

These are the kinds of questions political philosophers ask. *Political Philosophy* is the study of humans as they relate to one another and to government. The idea is that, once we understand what sort of beings humans are, we can know what sort of government best suits us. Political philosophers are also often concerned with economics because economics offers an empirical way to describe how humans interact with one another. In this chapter we'll investigate a few of the central issues surrounding these questions.

What Are Groups of Humans Like?

The primary issue in political philosophy—the question that must be answered before any other question is asked—is: What is the nature of humanity? For instance: Are we nasty and brutish? Are we generally good? Are we somewhere in between? Are we rational enough to make informed choices for ourselves? Are we free to make decisions independently of our genetics and upbringing?

If this question is not answered, we can't even begin to ask what sort of government would best suit humans. If humans are selfish and mean, and do not reason very well for themselves, we should choose a government that intervenes in people's lives on a regular basis, protects them, helps them, and makes decisions for them. We would want a "big" government, that is, a proactive government. A proactive government is involved in large areas of its citizens' lives. On the other hand, if people are generally pretty good to one another, or at least they can make rationally informed choices for themselves, then government could very easily get in the way, become an inconvenience, even burdensome. Therefore, we would want a "small" government, that is, a reactive government. A reactive government is involved only in very few areas of its citizens' lives, primarily for resolving disputes among them and for protecting them from outside invasion.

In most ancient cultures, individuals were considered merely one small part of a valuable group, and no individual was worth as much as the group. Individuals were considered only insofar as they were members of specific families, clans, or tribes, and could not make decisions for themselves. Based on this view of humanity, it makes sense why heavy-handed monarchies were the most common form of governement. In more recent times, governments have been more optimistic about their citizens. In the United States, the authors of the Constitution believed that individuals have rights even before they are members of a government, and that any government that is formed has a moral obligation to respect those rights. These two views of humanity make up the extreme ends of a continuum that influences how governments are best structured. These polar extremes are reflected in the two central philosophical theories of government: *Paternalism* and *Liberalism* (see Fig. 12-1).

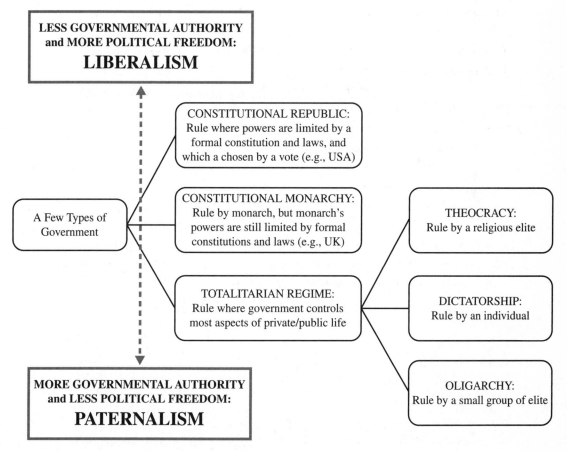

FIGURE 12-1• A continuum of individual—government relationships.

Paternal Governments

Paternalism (derived from the Latin, *pater* = "father") is the idea that the government, or some governmental body, is like a parent, while the governed are like children who need proper guidance and direction by the governed. Paternalism is motivated by a theory of human nature that says humans are either (a) rational but cruel, or (b) irrational. If humans are rational but cruel, they will behave only in ways that will help them achieve their goals regardless of harm or injustice to anyone else. If people are irrational, then they will not be able to understand what is best for them, or will not be able to obtain their best interests efficiently, thereby harming either themselves or others. A paternalistic government sets restrictions that prevent people from harming one another, intentionally or unintentionally. Censorship of various media—for example, games rated M for mature, violent books, pornographic movies—is one way a paternalistic government exercises control over its citizens. Taxation is another way.

Proponents of paternalism contend that some humans are fit to rule and the rest are fit only to be ruled. This sounds harsh, but it need not. Plato argued that people were born into one of three classes: the ruling class, whose souls were mixed with gold; the auxiliary class (of soldiers), whose souls were mixed with silver; and the producer class (peasants), whose souls were mixed with bronze or iron. You cannot help which class you are born into, and you will only be happy and successful when you live and work in your class. Therefore, Plato suggests a very large, very paternalistic government to help people find their appropriate class and live happily. Happy citizens have fewer reasons to hurt one another and to revolt against their government, therefore, according to Plato, a paternalistic government will be the most stable and best suited to humanity.

Unfortunately, we know from history that humans who are given power to rule other humans have been very bad at determining what will make others happy. No matter how often slave owners told slaves that they would be happier if they just did their work and kept their mouths shut, they just wouldn't listen (imagine that!).

BOX 12-1 Contemporary Paternalism

As of the writing of this book, China has censored or banned certain aspects of Google, various pornography sites, and other Internet groups such that people connected to the Internet in that country are unable to view many Web sites. This is a straightforward example of a paternalist regime attempting to look out for the interests of its citizens as if those citizens were children who cannot think for themselves.

Paternalistic ideas can also be found in the political writings of Aristotle and Thomas Hobbes (1588–1679). For Hobbes, people are self-centered, irrational jerks who need to have a strong government—an all-powerful Leviathan—to keep them scared enough not to harm one another. "Man is a wolf to man," claimed Hobbes, and their only hope is to enter into a contract with a governmental authority that will restrain their animal urges.

In addition, many Christian monarchs throughout the Middle Ages, Renaissance, and Enlightenment justified paternalism on the grounds that God appoints monarchs to rule sinful men, a view known as the *divine right of kings*. The view includes the idea that just as Adam was a kind of father of all of humanity (as is communicated in the Bible), so, too, the monarch of a society is the father-figure of its citizens, and this power is granted directly by God himself.

Liberal Governments

If you're almost offended by the thought of the government treating its citizens in this paternalistic way, then you probably have a much more optimistic view of humanity. You may believe that a less paternalistic, more liberal government best suits humans. *Liberalism* (derived from the Latin, *liber* = "free") is the idea that the government or governing body of a society is more like a referee than a parent. The referee sits outside the field of play, so to speak, and only intervenes in the private or public affairs of the governed when its necessary to "call a foul" for unfair play or adjudicate conflict. Such a government "lets things be," or is *laissez-faire* about governing. Note, too, that a referee does not have absolute power in a game, and can be challenged or even ousted by the players. If the referee cannot be ousted, the players are free to leave the field of play. So too, in social situations like a constitutional republic or constitutional monarchy, the citizens are free (hence, the *liber* part) to change the governing body without fear of harm or death. Thus, the government doesn't have the kind of power over citizens as one would find in a paternalist regime.

> ## BOX 12-2 Free to Change the Rules
>
> This freedom to alter governments is found in the U.S.'s Declaration of Independence from July 4, 1776: "We hold these truths to be self-evident, that all men are created equal, that they are endowed by their Creator with certain unalienable Rights. That among these are Life, Liberty and the pursuit of Happiness. That to secure these rights, Governments are instituted among Men, deriving their just powers from the consent

of the governed. *That whenever any form of Government becomes destructive of these ends, it is the Right of the People to alter or to abolish it, and to institute new Government, laying its foundation on such principles and organizing its powers in such form, as to them shall seem most likely to effect their Safety and Happiness* (italics added)..."

Liberal political ideas can be traced back to some of the philosophers we've already discussed in this book, including John Locke, Immanuel Kant, and John Stuart Mill. For Locke, people are inherently rational, and they are rational enough to govern themselves without a government. Locke's ideas, even some of his phrases, are present in the constitutions of the U.K. and the U.S., and the U.S.'s Declaration of Independence.

Mill also argues that people are inherently rational, and he contends that their primary goal is the pursuit of happiness. And who knows better what makes *you* happy than *you*? Thus, the freer you are to pursue your own happiness, the more likely you are to achieve it. Therefore, freedom of expression and ideas were of paramount importance for Mill. Mill also was influential in defending the full rights, privileges, and respect for women in a period known for its patriarchal tendencies.

Of course, with all the freedom, some people wondered how a liberal government could last without devolving into a chaos of lawlessness. What would keep most people from deceiving and coercing one another every chance they get? In response, Scottish philosopher Adam Smith (1723–1790) explained how economic *markets* actually regulate themselves. Smith argued that if people pursue their interests, producers will respond by producing what people want. If people stop wanting what is produced, producers will produce something else. He calls this principle of self-regulation the *invisible hand* of the marketplace. As long as trade remains free from government coercion, then by this invisible hand, producers will make money on the products they sell and then have money to buy the products that others sell.

The idea here is that the government should just allow people to pursue their own self-interests—in business transactions, and elsewhere—and the interactions of people will work themselves out to the benefit of all in the long run. The government should just let things be (*laissez-faire*) and not overburden citizens with laws. The combining forces of the ideas of Locke, Mill, and Smith form the strong basis for what is known as *classical liberalism* in economics and political philosophy.

None of this implies that we do not need a government. Classical liberals are adamant that government is necessary to react to corruption. If someone has been deceived or coerced to give up property they would not have given up, the government offers a means of redress. Citizens in dispute can call on law enforcement and courts to resolve disagreements and seek restitution. Government is also necessary to train and deploy military forces against outside attack. For classical liberals, the government is the watchman who sits on the tower to guard against corruption and invasion. Because of this, classical liberals often use the term "watchman state" to describe the role of government.

Freedom and Ownership

We have noted that if humans are inherently mean or simply irrational and unable to make decisions in their best interests, some version of paternalism is probably the best type of government. On the other hand, if humans pursue their self-interests rationally, then some version of liberal government is probably the best. But these are extreme versions of a continuum of positions. How do people decide how much freedom is enough? And is a government *ever* justified in restricting one person's freedoms for the sake of another's?

To answer these questions we need to know a little more about the type of *freedom* under consideration. We spoke a bit about freedom in Chapter 8. There we defined external freedom, political freedom, and internal freedom. *External freedom* is a kind of freedom whereby a person can do what he wishes without *external* obstacles or blockage. Every political philosopher would argue that this kind of freedom should be regulated and restricted to some degree in order to prevent people from harming one another in their individual pursuits. In liberal societies people generally have more external freedom, while in paternalistic societies people generally have less external freedom.

Political freedom is a type of external freedom. If you do not have political freedom, some external force, namely, government, is restricting what you are free to do. Political freedom manifests itself when the citizens of a society are free to vote, effectively appeal against an abusive government, actively engage in the law-making process, and trade freely and fairly with one another. This kind of freedom was pursued by William Wallace and the Scots in opposition to King Edward I of England at the end of the thirteenth century, and by the American colonies in opposition to King George III of England at the end of the eighteenth century. So, looking back at Fig. 12-1, we can see that less government authority, more political freedom, and liberalism go hand-in-hand, while more government authority, less political freedom, and paternalism go hand-in-hand.

Why would a liberal government have more political freedom than a paternalistic government? First, as we have seen, people who defend liberalism argue that people do not need very much regulation—they are beings capable of responding to reason. And second, they also argue that rational beings have a right to what they own, and that it would be immoral for a governmental authority to restrict what you are allowed to do with what is rightfully yours. We will discuss the concept of "rights" further on in this chapter, but for now we will focus on "ownership."

What does it mean to *own* something? John Locke, one of the earliest proponents of liberal government, argues that you are born owning your body and the work of your hands, and by mixing your work with property, it becomes yours:

> Though the earth and all inferior creatures be common to all men, yet every man has a *property* in his own *person*. This nobody has any right to but himself. The *labor* of his body and the *work* of his hands, we may say, are properly his. Whatsoever, then, he removes out of the state that nature hath provided and left it in, he hath mixed his labour with it, and joined to it something that is his own, and thereby makes it his property (*The Second Treatise of Government*, ch. V, section 27, emphasis his).

Property is obtained by working for it. This explains why an employer is obligated to pay you for working in her field or factory. You have mixed your labor with property they mixed their labor with, and therefore, you are owed something for your time and energy. This also explains why charity work is a good thing, since you are freely giving something of value, namely, your time and energy mixed with the property of others, that is, you are giving your property.

Does this mean you can sneak onto someone's property, farm for a few hours, and then demand payment? No. Since the farmer did not consent to your mixing your work with his property, you have stolen from him, and whatever work you put in must be considered just like charity—you freely gave it away.

So, why would anyone restrict what I can do with my own property (by taxing it, or capping how much money I can make)? Paternalist governments restrict freedoms much more than liberal governments for one of two reasons. They either deny the existence of personal property; that is, they deny that when you are born you belong to yourself—you are a ward of the state and exist for others—or they deny the possibility of fair exchange of property. If you are not your own person with individual rights prior to the existence of a government, then the government establishes your rights and provides for your needs. You belong to the government. On the other hand, even if you have rights independently of government, if all trade is tainted by corruption, then even if your body and work are your own, nothing you mix your work with will be rightfully yours because you are mixing your work with stolen goods. How might corruption taint trade? Just think of how your country was founded. Unless you live in Liberia (which was founded by freed American slaves), your country was probably founded by a war motivated by political expansion—so your land was stolen from someone. This may not be the case in every situation, but it may be that enough corruption exists to taint all future trade. All trade is an exchange of stolen goods, thus you do not rightfully own even that which you worked for.

Of course, this second motivation for paternalism is only plausible if there is no way to eliminate unfairness from the system. If unfairness can be redeemed by restitution or enaging in honest trade, or if unfairness is eliminated as the wronged parties die out or the society forgets the injustices, then a system of fair trade remains possible. And, in fact, liberal governments argue that it is precisely because corruption is possible that we need government to punish those who act unfairly.

BOX 12-3 CRITIQUE IT!

It is hard to justify a strongly paternalistic government, especially if one comes from a Western perspective where it seems obvious that people should be free to alter or abolish their government. Also, the kinds of horrors, injustices, and other abuses that history has shown to take place in totalitarian regimes (e.g., Nazi Germany, Stalin's U.S.S.R., Mao Zedung's People's Republic of China) certainly does not put paternalism in a good light. However, liberalism must be kept in check, too, because there are many who think that their external freedom allows them to commit injustices and other abuses. Consider the kinds of abuses that resulted from the Industrial Revolution in places like the U.K. and U.S. by the end of the nineteenth century where people (including children) were literally worked to death, or lost limbs, or picked up diseases in Ebenezer Scrooge-like working and living conditions.

What people often do not see are the places where paternalism creeps in unobtrusively: more taxes for government programs, bans on smoking in private businesses, fines for not recycling. All of these increases in governmental regulation are justified on the grounds that we need to look out for one another or that we should all care about the environment. But these justifications are irrelevant. The real question is: Is it the government's job to *force* me to look out for someone else or to care about the environment?

Law and Morality

Law refers to a rule or guideline that a person should follow in acting, which is enforced through some power that is able to enforce it. Laws either protect or restrict political freedom. The hope is that laws are the guidelines that help ensure that the society functions optimally and in peace. Usually a law is phrased as a curtailing of someone's political freedom, so that the most number of people can coexist most optimally. For example, in most states in the U.S., there is a law stating something like: "It is illegal to yell 'fire' in a crowded area when, in fact, there is no fire; as this could cause needless harm to people who are scrambling to get out of the area (someone could get trampled upon, for example)." Often—though not always—a societal law is put in place *precisely because of* a moral law. In other words, the societal law is based on the moral law. Some obvious examples are the shown in Fig. 12-2.

Moral Law, which is basis for: →	Societal Law
"One should/ought not murder"… and this is the basis for →	It's against the (societal) law in society X to murder, it's declared in writing, and you will be tried in court, and punished with jail time, fines, and/or the death penalty if you do murder…
"One should not steal"… and this is the basis for →	It's against the law in society Y to steal, and you will be punished accordingly…
"One should not commit fraud"… and this is the basis for →	It's against the law to commit fraud in business transactions…
"One ought not lie"… and this is the basis for →	It's against the law to lie in an official court case…

FIGURE 12-2 · Moral law vs. societal law.

Moral principles are typically justified according to rational principles, such as those of Immanuel Kant, Mill, or Aristotle, or religious principles (e.g., God has given you the right to life, liberty, and the pursuit of happiness). And, interestingly enough, each of these positions (as well as any moral position) will be influenced in large part by how a particular society views what exists in the world, namely, the *metaphysical* viewpoint, worldview, or outlook on reality.

For example, in a theocratic society (a religiously-based political system; *theos* is Greek for "god") there may be a societal law that says a man must have a certain appearance when out in public. This societal law may be based on a moral law that is derived from a religious text. And, the religious text is significant because in a theocracy, people believe in a god who has influenced the writing of the scripture and the kinds of moral laws found in it. We could add this theocratic metaphysical outlook layer to the table above, so that in theocratic society X, for example (Fig. 12-3):

Metaphysical Outlook—What Exists, which is basis for →	Moral Law, which is basis for →	Societal Law
For theocratic society X, in reality there exists people, the rest of the universe, and a god. The god is very, very important in this society, the god co-wrote scripture, and in the scripture this god tells us that "Thou shalt not murder"… and this is the basis for →	"One should/ought not murder"… and this is the basis for →	It's against the (societal) law in theocratic society X to murder, it's declared in writing, and you will be tried in court, and punished with jail time, fines, and/or the death penalty if you do murder…
This god also told us in the scripture that "Thou shalt not steal"… and this is the basis for →	"One should not steal"… and this is the basis for →	It's against the law in society Y to steal, and you will be punished accordingly, possibly by losing your hand…
This god also told us in the scripture that "Thou shalt not commit adultery"… and this is the basis for →	"One should not commit adultery"… and this is the basis for →	It's against the law to commit adultery and you may be stoned to death…

FIGURE 12-3 · The relationship between society and metaphysics in a theocracy.

Alternatively, if a society explicitly rejects all religious foundations for morality, we might say this is an atheistic society. In an *atheistic society* moral laws might derive from a secular utilitarian philosophy whereby the moral decision is always the one that maximizes benefit for the majority, and their laws would reflect this moral theory. Some examples of societal laws in an atheistic society might be the following (Fig. 12-4):

Metaphysical Outlook—What Exists, which is basis for →	Moral Law, which is basis for →	Societal Law
For atheistic society Y, in reality there exists people, the rest of the universe, and nothing else. Humans figure out reality, including what is moral or not, and they have determined that a utilitarian position is rational and good to hold to… and this is the basis for →	"If one's end of life is causing misery to that person and to others, and is a financial drain on the community as a whole, then one should/ought to end one's life, or at least have that as an option"… and this is the basis for →	The societal law in atheistic society Y allows for suicide as well as physician-assisted suicide…
For atheistic society Y, there is a strict separation of religious matters from state matters… and this is the basis for →	"One ought to use state resources on state matters"… and this is the basis for →	It's against the law in atheistic society Y to use state resources for religious matters…

FIGURE 12-4 · The relationship between law and metaphysics in an atheist society.

As a real-life example, Rob was once in a conversation with someone who made the following claims—all in one breath, without missing a beat—and we can see the metaphysical outlook → moral law → societal law interconnection most readily:

Homosexuals are disordered by nature [metaphysical outlook], and their disorders lead them to commit immoral acts with their private parts [moral law]. They should not be allowed to marry or have any rights in a society [societal law], really, because they're sick [metaphysical outlook, again] and they act on that sickness [moral law, again].

What this person had to say actually echoes a common view of homosexuality in relation to a social issue such as gay and straight marriage. Thus, if people who believe this were the majority in a democratic society, it would be illegal for homosexuals to get married.

The point of this example is to show that the relationship between an individual and her government is more complex than merely considerations of rights and human nature, even if it shouldn't be. Often our perceptions of reality ultimately dictate what moral and social laws are demanded and enacted.

Individual Rights vs. The Common Good

Now let's take a step back and ask how we might mitigate these metaphysical assumptions. How do we determine whether our society should be more religious or more secular? This is a moral consideration, which leads us to ask about the moral relationship between individuals and their government. This is usually expressed in terms of "rights."

Ever since humans started settling down in communities, their interaction with one another has involved a constant tension between (a) the pleasures, goods, interests, and benefits of a singular person (the individual), on one hand, and (b) the pleasures, good, interests, and benefits of the entire community, on the other. And because some people want more than they have a right to, this tension has sometimes been quite intense. Thus, societal laws emerged as a way to adjudicate between individual interests—individual "rights"—and the interests of the community—the so-called *common good*—in a fair and just manner.

Rights are acts that I perform that do not obligate anyone else. For instance, I can think without requiring anything from anyone. I can walk my property, and no one has to lift a finger. I can raise my arms, etc. Now, the moment I try to do something that requires something of someone else, I am infringing on their rights—unless, of course, I make an agreement with them. For instance, if my neighbor grows potatoes and I don't, I can't just take one of his potatoes for myself—this would require him to give up something he worked for without consent. But, if I raise chickens and he doesn't, I could offer him one chicken for, say, a half bushel of potatoes. If he perceives that as a good deal, then we can agree to both benefit from the transaction; I get potatoes I didn't have and he gets a chicken he didn't have. No one's rights have been violated since no one was forced to do something he didn't consent to.

Some philosophers take this rather minimal, "negative" account of rights and add to it, arguing that we are entitled to quite a bit more than only what doesn't require anything of anyone else. They argue there are "positive" rights. I have a right to take some things from you against your will—like your income to pay for my healthcare, your income to pay for a war I don't like, your income to pay for welfare and WIC of people I don't know, etc. These extra burdens on others is justified on the grounds that it benefits the "common good." Strictly speaking, if we really have negative rights, this is stealing, even if you agree to pay it, since the alternative includes fines and imprisonment.

Interestingly, most people could care less that the money they earn is being used for people they've never met, and may not even like. For example, there are laws in many societies making citizens pay a public income tax related to their earnings at work and a public sales tax related to anything that they purchase. There are many other public taxes, too. These public taxes are given to government officials and used for the community interests of society on things like education, electricity, clean water, and many other interests. Of course, the individuals in any society don't usually jump for joy knowing that they have to pay taxes (which highlights the fact that individuals are concerned mostly with themselves), but most people do pay the taxes because they are aware of the fact that there these so-called "common goods" that need to be addressed and satisfied.

To know whether the "common good" is an adequate justification for taxation, we need to know something about these rights we have. Where did they come from? How strict are they? Many people will ground rights in a god of some kind, as is the case for the drafters of the U.S. Declaration of Independence back in 1776. Just before letting us know that "life, liberty, and the pursuit of happiness" are three examples of rights that U.S. folks have, we are told: "We hold these truths to be self-evident, that all men are created equal, *that they are endowed by their Creator* (italics added) with certain unalienable Rights.

That among these are Life, Liberty and the pursuit of Happiness…" There is no doubt that if you can convince people that some all-powerful god is the basis for a right, then the right will stick for sure!

BOX 12-4 CRITIQUE IT!

Invoking a "god-given right" is questionable because:

- You have to be specific about which god you're talking about. But then, when you are, you have the problem that not everyone believes in that god. Hence, they won't accept the reason you are giving for the right in question.

- Many people who appeal to religion, when pressed, could offer wholly secular reasons for the right in question. If this is possible, why invoke god as as a justification in the first place?

Another basis that is appealed to for individual rights has to do with the simple fact that an adult human being is just plain sacred, and worthy of dignity and respect *by virtue of* being rational, autonomous, free, and fully responsible. Thus, rationality, autonomy, freedom, and full responsibility ground a person's rights to certain things. In the history of Western philosophy, this view can be traced to thinkers such as Immanuel Kant (1724–1804) and John Stuart Mill (1806–1873), both of whom we saw in the chapters dealing with moral philosophy.

For Kant, given that persons are conscious, rational beings, capable of making their own free and informed decisions, they must be considered as *autonomous* beings. The word *autonomy* is made up of two Greek words meaning "self" (*auto*) and "law" (*nomos*), highlighting the fact that a rational person is "self-ruling" and his or her own informed decisions should be respected. Further, by virtue of their conscious rational capacities, humans have an innate worth, dignity, or respectability which automatically endows them with certain basic rights.

Mill thinks that many people are rational and worthy of rights in much the same way Kant does; however, he is willing to grant rights even to those folks who are not as rational and autonomous as we would like them to be in a society. In fact, as we saw in Chapter 10, Mill espouses utilitarianism and this moral theory is based on the fact that all individuals in some social setting must be treated equally and as possessing equal rights to the general goods of a society. Otherwise, there could not be fair decision-making with respect to maximizing the general pleasures/goods/benefits in some social setting.

We said that tensions arise in societies between individual pursuits and the common good. We can now more accurately characterize those individual pursuits as *rights* that an individual has in a social setting. Now, when a right is violated, inhibited, or negated in some way, the person usually claims something like "That's not fair," which is an accurate statement, and which is really a claim about justice. Regardless of their source, if we really have negative rights, then no amount of common good can outweigh the unfairness of taking from us against our will.

We now return briefly to our discussion of the influence of metaphysics. The nature of a person's rights will determine the extent to which a government can motivate political action on the basis of metaphysics. For example, in a country that has established "Sharia," or Islamic Law, individual rights are subsumed under the culture's respect for the teachings of Islam. However, if individual rights were recognized as primary, Islamic practice would be permitted, though it would not coerce people who were not of Islam to participate in the dictates of Islam. Historically speaking, governments that are more liberal tend to be more diverse, more tolerant, and more peaceful than more paternalistic governments. Liberal governments tend not to be blatantly religious or blatantly secular, and thus the metaphysical influence on policy and law is mitigated.

Still Struggling

You may be wondering how these very extreme views of paternalism and liberalism are relevant to current political climates. It is important to note that there has never been a wholly paternalistic or wholly liberal political system. Often, societies with generally "free" economic markets (where businesses are generally free to produce and sell their goods independently of the government) are heavily regulated by the government to prevent discrimination, monopolies, and fraud. Similarly, even societies in which the government owns the means of production can be driven by market forces of supply and demand with the goal of making more profit. We see these opposing perspectives every time lawmakers review statutes: for example, someone may propose a new law against texting and driving (paternalistic), while someone else suggests fewer restrictions on the adoption procedure (liberal); or someone may propose a higher sales tax to support the homeless (paternalistic), while someone else proposes lowering taxes for the middle class (liberal).

Since it is unlikely that there will ever be a purely liberal or purely paternal government, the most pressing question we face is how much should the government be involved in our lives, and what rational principles establish this limit?

Justice

Justice is often thought of as the *fair* or *equal treatment* of members of society by their government. To be sure, fairness and equality are not the same thing. Many economists argue that even in a fair economic market, there will be vast inequalities of income. This makes sense because some people work harder than others, and some people invent valuable products that seems to entitle them to the value that other people express for those products by paying for them. But, whereas some people argue that fairness should be the measure of justice, others argue that equality should take priority.

Those who argue that equality should be the measure of justice often defend a type of justice known as "distributive," which suggests that we guarantee that everyone receives an appropriate amount of good and resources within an economy. Opponents suggest that it simply means taking from those who have a lot and giving to those who have less, regardless of who deserves what. Supporters, on the other hand, argue that, in virtue of being human (or citizens, or moral beings), we have all have rights to certain things, regardless of who we have to take them from to provide them. Because of this, distributive justice often implies proactive regulation, that is, it seeks to prevent harms in a society by restricting freedoms such as speech and trade.

Those who argue that fairness should be the measure of justice, often promote the priority of "procedural" justice over distributive justice. Procedural justice is a view of justice that suggests that we should allow people to interact freely with one another (since individuals know best what is in their interests), and that the government should only intervene when a wrong has been committed. Distributive justice is reactive in that it does not seek to prevent harms before they occur.

- *Distributive justice* refers to the fair distribution of goods, privileges, and services, within an economy. Distributive justice often requires heavy regulation of practices in attempt to prevent harm. For example, in the U.S., all citizens are afforded Social Security at a certain age, and this is based on the concept of distributive justice.

- *Procedural justice* refers to the fair handling of harms committed in a society. Procedural justice allows fair practices to dictate the demographics of a society and only steps in to right wrongs that are committed—it does not always prescribe the prevention harms (although speed limits and licenses for drivers and professionals like doctors are exceptions). For example, in a criminal case where an offender has not paid government taxes, the offender must at least pay the taxes owed to the government and probably be penalized in other ways, too. Or, in a case where someone causes the accidental death of a child, the person should be held financially responsible for the death.

Distributive and procedural justice are not mutually exclusive. They may operate alongside one another in a political system. Indeed, many political philosophers (regardless of whether they are conservative or liberal) would argue that both are necessary to ensure the greatest amount of goods and freedoms within a society. However, the limits of distributive justice are often unclear in a political system, and increases in taxes and regulation inteded to make some people better off are often controversial. Opponents argue that distributive justice is simply a way to reintroduce "paternalism."

The following are examples of situations where the rights of the individual are put into conflict with the common good of the society and questions concerning distributive justice and procedural justice emerge:

- There have been many times where people in an American city have been forced to sell their homes to that city and move so that their homes may be leveled to make way for a new stadium, or a new casino, or a new facility that will generate much revenue for the city. Granted, in these circumstances people are almost always given some payment for their homes (though rarely what they're worth), but there are still issues of rights and justice that come to the fore:

  - On one hand, an individual can claim that her basic right to own a home freely in any place in the U.S., free from coercion, is violated or negated. After all, they are being forced to move.

  - On the other hand, the common good of the city is served because the new stadium, casino, or facility will bring in lots of revenue that is good for the city in terms of more jobs for people, more people spending money in the city, and more money for the city that can be used for city needs like education, tourism, etc.

- The welfare system in some societies is controversial for many people. There are those who claim that the goods that are afforded an individual in the welfare system are consistent with basic rights. Not everyone is born into decent circumstances; so there are those who need financial assistance from the government with food, housing, education, and other needs. At the same time, there are others who claim that the welfare system is a major drain on a nation's economy and that it perpetuates itself since folks get on welfare, abuse it, and never come off. Unfortunately, there is often more rhetoric than argument surrounding these issues: The "hard-earned" taxes from "hard-working" folks go toward these welfare "leeches." This is obviously a question of the extent of distributive justice, where the tension between individual rights and the common good runs high.

- Gun control is another hot-button issue that tests the limits of rights, common good, and justice, especially for Americans.

  - On one hand, an American individual can claim that it is his basic right to own a gun based on (a certain reading of) the Constitution, as well as for basic self-defense. It would not be fair to deny a rational, responsible adult the ability to own a gun.

  - On the other hand, the common good of the nation may be put to the test if many folks are allowed to carry firearms. Here, too, rhetoric is common: We don't want the Wild Wild West again, do we?

Still Struggling

Determining where we stand on political issues can be frustrating. There is not much information available about a candidate's actual beliefs about an issue, there is no way to tell exactly what he or she will do when in office, and then, regardless of who wins, our circumstances rarely change. So, why care? Caring becomes important on an individual level when you are trying to decide how to vote for local changes. More sales tax? More small business taxes? (It never seems to be less!) Do I join a union? These decisions reflect an understanding of how important we consider our rights and who we allow to regulate them. Making these decisions at an individual and local political level can be satisfying and empowering. And perhaps the more time we spend with these questions, the better educated voters will be, so that philosophically-informed policies will be represented at the national level.

Can't We Just All Get Along?

Issues in politics and political philosophy are some of the most emotionally charged, no doubt, and we all need to take a step back from our political positions, count to ten, then engage one another rationally. It's easier said than done, for most, that's for sure. Societies with more liberal governments usually afford their citizens more of an opportunity to discuss matters, vent, and agree to disagree, while those citizens in paternal states often have fewer opportunities for open dialogue. Hopefully a book like this will make it onto bookstands in paternalistic countries anyway!

QUIZ

1. If humans are _____, we should choose a government that can intervene in people's lives on a regular basis, protect them, help them, and make decisions for them.

    A. nice to each other
    B. fully rational
    C. selfish and mean
    D. conscious

2. What's one principle way that a paternalistic government exercises its control over its citizens that we specifically mentioned in this chapter?

    A. Putting citizens in jail
    B. Censorship
    C. Neither of the above
    D. Both of the above

3. The Latin *pater* in English means "free."

    A. True
    B. False

4. Whose idea was the *invisible hand*?

    A. Locke
    B. Smith
    C. Mill
    D. Jones

5. Ebenezer Scrooge, the infamous miserly character from the Charles Dickens classic *A Christmas Carol*, would likely have favored a government of this type:

    A. classical liberalism.
    B. paternalism.
    C. totalitarianism.
    D. none of the above.

6. In this chapter, it was noted that moral laws ground a metaphysical outlook.

    A. True
    B. False

7. The fact that I have a right to take some things from you against your will—like your income to pay for my healthcare and your income to pay for welfare—is what kind of right?

    A. Negative right
    B. Neutral right
    C. Negligible right
    D. None of the above

8. In a case where someone causes the accidental death of a child, _____ justice would demand that the person should be held financially responsible for the death of the child, as well as even do jail time.
   A. vigilantive
   B. distributive
   C. procedural
   D. none of the above

9. Liberal governments tend not to be blatantly religious or blatantly secular, and thus the metaphysical influence on policy and law is _____.
   A. heightened
   B. slightly less heightened
   C. exacerbated to the extreme
   D. mitigated

10. Historically speaking, governments that are more paternal tend to be more diverse, more tolerant, and more peaceful than more liberal governments.
    A. True
    B. False

Final Exam

1. Philosophy is the systematic study of reality using good reasoning in order to clarify difficult questions, solve significant problems, and _____.

   A. enrich human lives

   B. question everything

   C. use scientific reasoning

   D. gather facts about everything

2. Reality includes those things and events that make up the world and claims about them.

   A. True

   B. False

3. Philosophers are primarily interested in _____.

   A. subjective reality

   B. people's perceptions

   C. obeying the laws of a society

   D. objective reality

4. A primary difference between Eastern and Western approaches to philosophy is that Western philosophers do firmly distinguish their broadly religious perspectives from their philosophical investigations.

   A. True

   B. False

5. Prior to Thales' philosophical description of the world, most discussions of reality were _____.

    A. purely rhetorical in nature

    B. accurate and meaningful

    C. wholly pragmatic

    D. pragmatic or religious

6. A theoretical philosophy is what you might write to explain how you teach, or how you do business, or how you view the world.

    A. True

    B. False

7. _____ is the study of knowledge.

    A. Epistemology

    B. Episteme

    C. Logical Atomism

    D. Gnosticism

8. Semantics deals with the structure of claims and how this structure can be combined with logical operators (such as: and; or; if..., then...; not; if and only if) to allow us to infer new claims.

    A. True

    B. False

9. Normative claims are claims

    A. about normal actions.

    B. about the ways things are.

    C. about the way things ought to be.

    D. about actions that are out of the ordinary.

10. In English, the Greek word *arche* means

    A. natural law.

    B. rational guidance.

    C. first principle.

    D. guiding laws.

11. Who predicted the solar eclipse of 585 BCE?

    A. Xenophanes

    B. Socrates

    C. Plato

    D. Thales

12. A claim does not express a state of affairs, but rather leads to more questions.

    A. True

    B. False

13. Possibly, necessarily, and impossibly are

    A. operators.

    B. modal operators.

    C. quantifiers.

    D. indexicals.

14. By definition, claims are about reality, and are therefore

    A. either true or false.

    B. either partially subjective or wholly subjective.

    C. partially subjective

    D. wholly subjective.

15. The premise is supported by the conclusion in an argument.

    A. True

    B. False

16. Arguments can only help us evaluate some claims; there must be some claims we take as

    A. certain.

    B. true.

    C. basic.

    D. perceptive.

17. **All deductive arguments are valid arguments.**

    A. True

    B. False

18. **Seeing an object as red and round, smelling it as an apple, and feeling it as smooth and waxy are all forms of**

    A. indirect evidence.

    B. mediated evidence.

    C. direct evidence.

    D. vaguely sufficient evidence.

19. **_____ are the least controversial sources of experiential evidence.**

    A. Sense experience and intuitive awareness

    B. Scientific experiments

    C. Direct evidence and indirect evidence

    D. Sense experience and scientific experiments

20. **Testimony is not a form of non-experiential evidence.**

    A. True

    B. False

21. **Rational intuition is not a form of non-experiential evidence.**

    A. True

    B. False

22. **In _____, an argument is constructed that has all the same elements as the argument under consideration *except* the act in question.**

    A. the Technique of Variant Cases

    B. the Technique of Invariant Cases

    C. the Variant Technique of Cases

    D. the Case of Invariant Technique

23. **Which one of these is not a theoretical virtue concerning explanation?**

    A. Simplicity

    B. Conservatism

    C. Explanatory scope

    D. Assessable claim-recognition

24. **The following is an example of what kind of fallacy? When President Bill Clinton cheated on his wife with an intern, many of his political opponents argued that this character flaw undermined his ability to perform his duties as president.**

    A. Appeal to the people

    B. Appeal to snobbery

    C. Appeal to vanity

    D. Appeal to the person

25. **The following is an example of what kind of fallacy? The *Journal of Paranormal Research* regularly publishes articles by people claiming to be experts in paranormal psychology. So, paranormal phenomena exist.**

    A. Appeal to the people

    B. Appeal to inappropriate authority

    C. Appeal to vanity

    D. Appeal to the person

26. **_____ skepticism is the view that humans do not know whether they have or can have knowledge.**

    A. Apodictic

    B. Platonic

    C. Pyrrhonian

    D. Globalized

27. **Who is the following claim wrongly attributed to: "All I know is that I know nothing?"**

    A. Thales

    B. Plato

    C. Socrates

    D. Kant

28. Descartes was a global skeptic.

    A. True

    B. False

29. Which of the following was not an argument for skepticism spoken about in this book?

    A. The cave argument

    B. The argument from reliable sense experience

    C. The dream argument

    D. The evil genius argument

30. The view that "I am the only thing that exists" is called

    A. solitary.

    B. solo-performance.

    C. solipsism.

    D. sophistication.

31. A claim is _____ if it does not have a precise meaning.

    A. equivocal

    B. vague

    C. ambiguous

    D. false

32. In order to reach the conclusion that most of his beliefs are reliable, Descartes would have to know that God desires that our senses accurately represent reality to us. This is known as

    A. Descartes' Law.

    B. the Argument from Circularity.

    C. the Cartesian Circle.

    D. none of the above.

33. Who defends the idea that our minds are passive recipients by citing examples from our mature adult perception and the perceptions of children?

    A. Descartes

    B. Plato

    C. Hume

    D. Locke

34. Who puts forward the *tabula rasa* view?

    A. Descartes

    B. Hume

    C. Locke

    D. Plato

35. Who puts forward the argument concerning the piece of wax, whereby knowledge is a function of the mind?

    A. Descartes

    B. Hume

    C. Locke

    D. Plato

36. Who not only relies heavily on Probability for making claims about the world outside our senses, but also admits that this is not sufficient for establishing an accurate and enduring scientific picture of the world?

    A. Locke

    B. Plato

    C. Descartes

    D. Hume

37. Hume's "matters of fact" include how our minds organize and understand the sensible world.

    A. True

    B. False

38. "Matters of fact" is to sensation, as "relations of ideas" is to

    A. perception.

    B. concepts.

    C. reflection.

    D. consciousness.

39. According to Hume, most of our claims about reality have little to do with what we directly experience.

    A. True

    B. False

40. Pragmatically, we cannot live without it; philosophically, it is irrational. To what is this referring here?

    A. Justification

    B. Belief-formation

    C. Induction

    D. Inferences

41. A *priori* is Latin for, "from within experience."

    A. True

    B. False

42. Who is the famous person who claimed that science doesn't need induction after all?

    A. Hume

    B. Kant

    C. Popper

    D. Salmon

43. Which is correct?

    A. Deductivism is a negation of falsifiability and corroboration.

    B. Deductivism is a combination of falsifiability and corroboration.

    C. Falsifiability is a combination of deductivism and corroboration.

    D. Falsifiability is a negation of deductivism and corroboration.

44. **For Hospers, without induction, the word "_____" would lose any semblance of meaning.**

    A. validity

    B. deduction

    C. falsifiability

    D. evidence

45. **Who interrupted Kant's dogmatic slumber?**

    A. Locke

    B. Aristotle

    C. Popper

    D. Hume

46. **According to Hume, the "I" is not available to sensory perception and any reasoning about the "I" makes the "I" the object of our reasoning, and therefore, we cannot, by definition, examine it as it operates subjectively. So, to know anything, we must begin with basic perception.**

    A. True

    B. False

47. **Kant divides the world into the "phenomenal" and the "noumenal."**

    A. True

    B. False

48. **For Kant, of the following, which form the necessary prerequisites for a complete understanding of morality?**

    A. Free will, phenomenal reality, and God

    B. Free will, God, and noumenal reality

    C. Free will, an eternal soul, and God

    D. An eternal soul, God, and noumenal reality

49. In 1963, a relatively unknown philosopher named _____, under pressure from the Administration of Wayne State University (where he was teaching at the time) to publish something, published a three-page paper, "Is Justified True Belief Knowledge?" that changed how philosophers approach epistemology.

    A. Edmund Husserl

    B. Edmund Gettier

    C. Wesley Salmon

    D. John Hospers

50. The first principles of any system cannot be proven within that system, on pain of

    A. individuation.

    B. unfounded speculation.

    C. circularity.

    D. irreconcilability.

51. The ability to solve a simple problem like using a stick to get at food just out of reach is a capacity of

    A. Perceptual mind.

    B. Conscious mind.

    C. Reasoning mind.

    D. none of the above.

52. The following is a capacity of conscious mind:

    A. The capacity to perceive.

    B. The ability to think about one as a self who is thinking.

    C. The ability to form beliefs about the past and future.

    D. All of the above.

53. No one can say *for sure, with absolute certainty* that an animal does or does not have a mind.

    A. True

    B. False

54. It's the actual size and weight of a brain that is the indicator of a mind's complexity.

    A. True

    B. False

55. Those who believe in the immortality of the soul, or reincarnation, are _____ because they think that the death of the body does not mean the death of the soul.

    A. substance dualists

    B. property dualists

    C. metaphysical intuitionists

    D. none of the above

56. _____ was the Father of Modern Philosophy.

    A. Locke

    B. St. Thomas Aquinas

    C. Descartes

    D. Kant

57. How can an immaterial mind (either as mental *substance* or mental *property*) "move" my brain and nervous system to cause it to raise my arm, for example? This is part of the problem of

    A. recollection.

    B. interaction.

    C. invasive-intuition.

    D. physiological causation.

58. Of the following, who was not a behaviorist?

    A. Ivan Pavlov

    B. B.F. Skinner

    C. John Watson

    D. Jamie Watson

59. Of the following, who was a psychoanalyst?

    A. Karl Jaspers

    B. Sigmund Freud

    C. Karl Popper

    D. B.F. Skinner

60. The elegance of Philosophical Behaviorism is that we do not need to refer to anything beyond _____ to explain what it means for Jones to "believe" something or "hope" something or "feel" something.

    A. mental states

    B. cognition

    C. behavioral dispositions

    D. intuitions

61. Eliminative materialists believe that any mental quality can be identified with an underlying neurobiological process and, hence, can be wholly identified with that neurobiological process, so that even our perception of mental states is _____.

    A. illusory

    B. necessary

    C. possibly important to our actions

    D. none of the above

62. For the functionalist, if in fact minds are like computer programs running on the hardware of the brain, then we should be able to develop highly complicated computer systems that will

    A. perform software functions.

    B. perform functions like that of computers.

    C. mimic minds.

    D. mimic computers.

63. Which of the following is not a criterion for personhood?

    A. Capacity to understand

    B. Capacity to speak a language

    C. Capacity to reason

    D. Capacity to read before puberty

64. Although someone might argue that very small children and the severely mentally handicapped are not persons, few would say that the *physically* handicapped are not persons. This raises the interesting matter of whether _____ is even necessary as a condition for personhood.

    A. a body

    B. a soul

    C. a mind

    D. none of the above

65. According to some philosophers, it seems possible to be friends with a brain in a jar.

    A. True

    B. False

66. Which of the following is not a mental state?

    A. A belief

    B. A feeling

    C. A music download

    D. Knowing your computer is downloading music

67. With respect to language, there is no need to draw a distinction between *communicating some information* and actually *engaging in a language*.

    A. True

    B. False

68. Of the following, what does not seem to count against the idea that the body is what identifies you as "you" throughout your life?

    A. Aging

    B. Coma

    C. Dementia

    D. Diabetes

69. "I can conceive of myself existing without a body" and "Conceivability entails possibility (that is, if I can conceive of some state of affairs, then that state of affairs is really possible—it is a genuine way the world could have been)." In the book, we attributed these two premises to:

    A. Locke

    B. Hume

    C. Descartes

    D. Kant

70. You are really just your collection of memories and past experiences or a causal chain of mental events, according to

    A. Locke.

    B. Hume.

    C. Descartes.

    D. Kant.

71. To be free in this sense entails the possibility of always being able to act other than one has acted if one had the opportunity to perform the action over again.

    A. External freedom

    B. Internal freedom

    C. Political freedom

    D. Alternate possibilities freedom

72. Here, the shadow of the *unchosen* action always attends every deliberate and chosen action.

    A. External freedom

    B. Internal freedom

    C. Political freedom

    D. Alternate possibilities freedom

73. **Democratic republics, such as the United States of America, embody this kind of freedom.**

    A. External freedom

    B. Internal freedom

    C. Political freedom

    D. Alternate possibilities freedom

74. **We don't hold the following people morally responsible for their actions: young children, the severely mentally ill, rational voting adults, the severely mentally disabled, persons with dementia.**

    A. True

    B. False

75. **_____ is the view that some event will happen in the future no matter *what happens* until then.**

    A. Fatalism

    B. Animism

    C. Materialism

    D. Determinism

76. **According to hard indeterminism, whereas every other kind of event in the universe has a history that someone with a _____ could trace and determine, _____ events are different in that they are not determined.**

    A. human mind, unaltered

    B. human mind, god-like

    C. god's-eye-view, human

    D. human mind, unaltered

77. **_____ argue that alternative possibilities freedom is the freedom that is necessary for moral responsibility. In fact, you can't be morally responsible for your actions even if you could not have done otherwise.**

    A. Compatibilists

    B. Fatalists

    C. Determinists

    D. None of the above

78. Philosophers of religion are careful to note that their burden is to prove that a god exists, not to presume it exists before any evidence has been offered.

    A. True

    B. False

79. Even if philosophers of religion establish that some divine being exists, they need not go any further and argue that their particular conception of God exists (e.g., the God of Judaism, the God of Christianity, or the God of Islam, etc.).

    A. True

    B. False

80. _____ would not accept religious experience as sufficient evidence of God's existence.

    A. An adventist

    B. A non-adventist

    C. An evidentialist

    D. A non-evidentialist

81. _____ is the view that non-evidentialism is true and that belief in God is best justified by whether it is in my best interests to believe in God.

    A. Atheism

    B. Evidentialism

    C. Agnosticism

    D. Prudentialism

82. Darwinian evolution is not a pure chance process.

    A. True

    B. False

83. God is that than which no greater can be thought in the

    A. teleological argument.

    B. evidentialist argument.

    C. ontological argument.

    D. Watchmaker argument.

84. Whereas evidentialist arguments attempt to justify belief that the claim, "God exists," is true or false, prudential arguments attempt to establish that you should believe (or not) *regardless of whether* the claim, "God exists," is true.

    A. True

    B. False

85. A *paternalistic government* can be defined as the authorized body (one person, or group of people) in a social setting who exercise the power and control over the laws, regulations, and funding concerning the social setting as a whole.

    A. True

    B. False

86. Plato argued that people were born into one of three classes: the ruling class, whose souls were mixed with gold, the auxiliary class (of soldiers), whose souls were mixed with silver, and the producer class (peasants), whose souls were mixed with bronze or _____.

    A. metal

    B. alloy

    C. platinum

    D. none of the above

87. The combining forces of the ideas of Locke, Mill, and Smith form the strong basis for what is known as _____ in economics and political philosophy.

    A. classical liberalism

    B. classical economics

    C. economical conservativism

    D. classical conservativism

88. Historically speaking, governments that are more liberal tend to be more diverse, more tolerant, and more peaceful than more paternalistic governments.

    A. True

    B. False

89. If you're in favor of a city taking over land for a city project and forcing people to sell their houses in order for the city to take over that land, then you're likely someone who would defend the priority of

    A. individual rights.

    B. the common good.

    C. democracy

    D. alternate possibilities freedom

90. _____ action is one that violates an obligation, that is, some rule that has determined the action to be immoral.

    A. A negligible.

    B. Anobligatory

    C. A totally permissible

    D. An impermissible

91. Telling the truth is an example of _____ action.

    A. an obligatory

    B. a mega-obligatory

    C. a permissible

    D. an impermissible

92. Ontology is the study of duty.

    A. True

    B. False

93. For all its internal problems, Kant's theory was an advance in moral thinking, and still forms the basis of much of contemporary talk about moral principles and blind justice, as well as individual rights and respect for personal decision-making.

    A. True

    B. False

94. It may be necessary to shoot down a hijacked plane to prevent it from crashing into a building where there are lots of people. This came up in the context of

   A. duty-based moral theory.

   B. virtue-based moral theory.

   C. consequence-based moral theory.

   D. none of the above.

95. Jesus' command in the Christian Scriptures to "love your enemies" is a clear case of

   A. duty-based moral theory.

   B. virtue-based moral theory.

   C. consequence-based moral theory.

   D. none of the above.

96. A person may be guilty without being blameworthy, for instance, if that person were mentally handicapped and, therefore, could not consider reasons for acting or not acting.

   A. True

   B. False

97. Moral agents must take responsibility for their actions, and if someone acts in such a way that will produce a being that will eventually be able to consider and act on moral reasons, she should not treat that being merely as a means to her own personal happiness. This idea is founded squarely in a

   A. duty-based moral theory.

   B. virtue-based moral theory.

   C. consequence-based moral theory.

   D. none of the above.

98. _____ seem to suggest that the best course of action for you depends on your psychology and which reasons for action will foster good actions in the future.

    A. Value-based moral theorists

    B. Consequence-based moral theorists

    C. Duty-based moral theorists

    D. None of the above

99. Not all theories are good theories. Even if they are internally consistent, they may be inconsistent with the best evidence we have about what is _____.

    A. valid and invalid

    B. valid

    C. right and wrong

    D. invalid

100. We are all tied to our culture in important ways, and our moral views are shaped partly by our cultures.

    A. True

    B. False

Answers to Quizzes and Final Exam

Chapter 1	Chapter 3	Chapter 5	Chapter 7
1. B	1. A	1. B	1. A
2. A	2. D	2. A	2. B
3. D	3. A	3. B	3. D
4. C	4. D	4. D	4. B
5. A	5. A	5. A	5. B
6. D	6. C	6. A	6. D
7. B	7. B	7. B	7. D
8. C	8. D	8. B	8. D
9. C	9. C	9. B	9. A
10. C	10. A	10. D	10. D

Chapter 2	Chapter 4	Chapter 6	Chapter 8
1. A	1. B	1. C	1. C
2. A	2. B	2. A	2. A
3. C	3. A	3. A	3. D
4. A	4. D	4. B	4. D
5. B	5. B	5. B	5. B
6. A	6. C	6. A	6. A
7. B	7. A	7. A	7. C
8. D	8. D	8. C	8. D
9. A	9. C	9. D	9. A
10. C	10. A	10. C	10. A

Chapter 9
1. C
2. B
3. B
4. C
5. A
6. D
7. B
8. B
9. C
10. B

Chapter 10
1. B
2. D
3. C
4. C
5. B
6. D
7. C
8. D
9. A
10. B

Chapter 11
1. B
2. D
3. C
4. A
5. D
6. B
7. A
8. B
9. C
10. A

Chapter 12
1. C
2. B

3. B
4. B
5. A
6. B
7. D
8. C
9. D
10. B

Final Exam
1. A
2. A
3. D
4. A
5. D
6. B
7. A
8. B
9. C
10. C
11. D
12. B
13. B
14. A
15. B
16. C
17. B
18. C
19. D
20. A
21. B
22. A
23. D
24. D
25. B
26. C
27. C
28. B

29. B
30. C
31. B
32. D
33. D
34. C
35. A
36. A
37. B
38. C
39. A
40. C
41. B
42. C
43. B
44. D
45. D
46. B
47. A
48. C
49. B
50. C
51. C
52. D
53. B
54. B
55. A
56. C
57. B
58. D
59. B
60. C
61. A
62. C
63. D
64. A
65. A
66. C

67. B
68. D
69. C
70. B
71. D
72. D
73. C
74. B
75. A
76. C
77. D
78. A
79. B
80. C
81. D
82. A
83. C
84. A
85. B
86. D
87. A
88. A
89. B
90. D
91. A
92. B
93. A
94. C
95. A
96. A
97. A
98. D
99. C
100. A

Suggested Reading

Chapter 1: Introduction

- William Irwin and David Kyle Johnson, eds. *Introducing Philosophy Through Pop Culture: From Socrates to South Park, Hume to House* (Malden, MA: Wiley-Blackwell, 2010).
- Nils Ch. Rauhut, *Readings on the Ultimate Questions* (New York: Pearson-Longman, 2010).
- Robert Paul Wolff, *Ten Great Works of Philosophy* (New York: Signet Classics, 2002).
- Louis Pojman and Lewis Vaughn, eds. *Classics of Philosophy* (New York: Oxford University Press, 2010).

Suggested Papers

(Links to online versions of each of these are given below.)

1. Epicurus, "Letter to Menoeceus." http://www.epicurus.net/en/menoeceus.html
2. Immanuel Kant, "What is Enlightenment?" http://philosophy.eserver.org/kant/what-is-enlightenment.txt
3. Bertrand Russell, chapters of *The Problems of Philosophy. http://www.ditext.com/russell/russell.html*
4. John Dewey, "The Need for a Recovery of Philosophy." http://www.brocku.ca/MeadProject/Dewey/Dewey_1917b.html

Chapter 2: Thinking Critically about Reality

- Jamie Carlin Watson and Robert Arp, *Critical Thinking: An Introduction to Reasoning Well* (London: Continuum Press, 2011).
- Anthony Weston, *A Rulebook for Arguments* (New York: Hackett, 2008).
- Darrell Huff, *How to Lie With Statistics* (New York: W. W. Norton, 1993).
- Theodore Schick and Lewis Vaughn, *How to Think About Weird Things* (Boston: McGraw-Hill, 2002).
- Robert Arp, "The Chewbacca Defense: A *South Park* Logic Lesson," in Robert Arp, ed., *South Park and Philosophy: You Know, I Learned Something Today*, pp. 40–54 (Malden, MA: Wiley-Blackwell, 2006).

Chapter 3: Knowledge and the Problem of Skepticism

- Sextus Empiricus, *Outlines of Pyrrhonism*, R.G. Bury, trans. (Amherst, NY: Prometheus Books, 1990).
- Réne Descartes, *Meditations on First Philosophy*, in Margaret Wilson, ed., *The Essential Descartes* (New York: Meridian, 1976).
- John Greco, ed. *The Oxford Handbook of Skepticism* (New York: Oxford University Press, 2008).
- Walter Sinnott-Armstrong, ed. *Pyrrhonian Skepticism* (New York: Oxford University Press, 2004).
- Keith DeRose and Ted Warfield, eds. *Skepticism: A Contemporary Reader* (New York: Oxford University Press, 1999).

Chapter 4: Responses to Skepticism

- Christopher Grau, "Bad Dreams, Evil Demons, and the Experience Machine: Philosophy and *The Matrix*," in Christopher Grau, ed., *Philosophers Explore The Matrix*, pp. 10–23 (New York: Oxford University Press, 2005).
- Linda Zagzebski, *On Epistemology* (Belmont, CA: Wadsworth, 2009).
- Bertrand Russell, "Idealism," in *The Problems of Philosophy* (New York: Barnes and Noble, 1912/2004).

- Louis Pojman, ed. *The Theory of Knowledge: Classic and Contemporary Readings* (Belmont, CA: Wadsworth, 2002).
- Robert Arp, "Frege, As-If Platonism, and Pragmatism." *Journal of Critical Realism* (2005) 4:1, 22–41.

Chapter 5: The Problem of Induction and the Development of Externalism

- John L. Pollock and Joseph Cruz, *Contemporary Theories of Knowledge* (Lanham, MD: Rowman and Littlefield, 1999).
- Alvin Plantinga, *Warrant: The Current Debate* (New York: Oxford University Press, 1993).
- Keith Lehrer, *The Theory of Knowledge* (Boulder, CO: Westview Press, 2000).
- Wesley Salmon, *The Foundations of Scientific Inference* (Pittsburgh: The University of Pittsburgh Press, 1966).
- Roger Scruton, *Kant: A Very Short Introduction* (New York: Oxford University Press, 2001).

Chapter 6: The Mind-Body Problem

- David Braddon-Mitchell and Frank Jackson, *Philosophy of Mind and Cognition* (Malden, MA: Blackwell Publishing, 2007).
- David Chalmers, *The Conscious Mind* (New York: Oxford University Press, 1996).
- William Alston, *The Emergent Self* (Ithaca, NY: Cornell University Press, 1999).
- Jaegwon Kim, *Mind in a Physical World* (Cambridge, MA: MIT Press, 2001).
- Robert Arp, "Consciousness and Awareness: Switched-On Rheostats." *Journal of Consciousness Studies* (2007) 14:3, 101–106.

Chapter 7: Personhood and Personal Identity over Time

- John Locke, "Of Identity and Diversity," *An Essay Concerning Human Understanding*, Book II, Chapter 27.

- Michael C. Rea, *Material Constitution: A Reader* (Lanham, MD: Rowmand & Littlefield, 1997).
- Daniel Dennett, *Brainstorms* (Montgomery, VT: Bradford Books, 1978).
- Derek Parfit, *Reasons and Persons* (Oxford: Oxford University Press, 1984).
- Robert Arp, "If Droids Could Think... Droids as Slaves and Persons," in Kevin Decker and Jason Eberl, eds., *Star Wars and Philosophy: More Powerful than You can Possibly Imagine*, pp. 120–131 (Peru, IL: Open Court Publishing Company, 2005).

Chapter 8: Freedom and Determinism

- Laura Waddell Ekstrom, *Agency and Responsibility* (Boulder, CO: Westview Press: 2001).
- Gary Watson, *Free Will* (New York: Oxford University Press, 2003).
- John Martin Fischer and Mark Ravizza, *Responsibility and Control: A Theory of Moral Responsibility* (Cambridge: Cambridge University Press, 1998).
- Randolph Clarke, *Libertarian Accounts of Free Will* (New York: Oxford University Press, 2003).
- Robert Arp, "Suarez and Filmer on Freedom," in Richard Corrigan and Mary Ferrell, eds., *Philosophical Frontiers: Essays and Emerging Thoughts*, pp. 39–66 (Gloucester: Progressive Frontiers Press, 2009).

Chapter 9: The Question of God's Existence

- William Lane Craig, *Reasonable Faith* (Wheaton, IL: Crossway Books, 2010).
- Michael Peterson, Michael Peterson, William Hasker, Bruce Reichenbach, and David Basinger, *Philosophy of Religion: Selected Readings* (New York: Oxford University Press, 1996).
- Nicholas Everitt, *The Non-Existence of God* (London: Routledge, 2004).
- William Lycan and George Schlesinger, "You Bet Your Life: Pascal's Wager Defended," in R. Douglas Geivett and Brenden Sweetman, eds., *Contemporary Perspectives on Religious Epistemology*, pp. 270–282. (New York: Oxford University Press, 1992).

- Jamie Watson, "The Beast in Me: Evil in Cash's Christian Worldview," in John Huss and David Werther, eds., *Johnny Cash and Philosophy: The Burning Ring of Truth*, pp. 227–238 (Peru, IL: Open Court Publishing Company, 2008).

Chapter 10: Moral Philosophy

- Robert Arp and Jamie Carlin Watson, *What's Good on TV? Learning Ethics through Television* (Malden, MA: Wiley-Blackwell, 2011).
- Louis Pojman, *Ethics: Discovering Right and Wrong* (Belmont, CA: Wadsworth, 1994).
- Simon Blackburn, *Ethics: A Very Short Introduction* (Oxford: Oxford University Press, 2003).
- Steven Cahn and Peter Markie, eds. *Ethics: History, Theory, and Contemporary Issues* (New York: Oxford University Press, 2008).
- Jamie Carlin Watson, "Holden Onto What's Right," in Keith Dromm and Heather Salter, eds., *The Catcher in the Rye and Philosophy* (Peru, IL: Open Court Publishing Company, 2011).

Chapter 11: Moral Decision Making

- Vincent Ryan Ruggiero, *Thinking Critically About Ethical Issues*, 7th ed. (Boston: McGraw-Hill, 2008).
- John Hospers, *Human Conduct: Problems of Ethics*, 2nd ed. (New York: Harcourt Brace Jovanovich, 1982).
- Barbara MacKinnon, "Ethics and Ethical Reasoning," Chapter 1, *Ethics: Theory and Contemporary Issues* (Australia: Wadsworth Cengage, 2009).
- Ronald A. Howard and Clinton D. Korver, *Ethics for the Real World: Creating a Personal Code to Guide Decisions in Work and Life* (Cambridge, MA: Harvard Business School Press, 2008).

Chapter 12: Justice, Rights, and Government

- Steven Cahn, ed. *Political Philosophy: The Essential Texts* (Oxford: Oxford University Press, 2010).

- Jonathan Wolff, *An Introduction to Political Philosophy* (Oxford: Oxford University Press, 2006).
- Steven Cahn, *Political Philosophy: A Very Short Introduction* (Oxford: Oxford University Press, 2003).
- Milton Freidman, *Capitalism and Freedom* (Chicago: University of Chicago Press, 2002).
- John Hospers, *Libertarianism: A Political Philosophy for Tomorrow* (Lincoln, NE: Author's Choice Press, 2007).

Index

A

a posteriori vs. *a priori* evidence, 94–95
abortion example, 226–229
Academic skepticism, 54
    plausibility of, 57
    vs. Pyrrhonism, 55, 57–58
    Socratic Method, 55
Achilles, paradox of, 14–15
ad hominem, 44
agnosticism, overview, 177–180
AI (artificial intelligence), 127–129, 146
Alexandria, Clement of, 13
al-Ghazali, 16, 193–194
allegiance relationships, 144
"The Allegory of the Cave," 58–59
Almanac, The Old Farmer's, 5–6
alternate possibilities freedom (APF). *See also* free will;
    principle of alternative possibilities (PAP)
    argument against, 168
    counterexample, 169
    vs. determinism, 163–164
    and moral responsibility, 159–161
    overview, 158–159
Alzheimer's disease, 145, 147
analogues, defined, 36
Anaximander, 16
"and" operator, 26
animals
    mental states of, 141–142
    minds of, 111–113
Anselm of Canterbury, 187–189
anthropomorphism, explained, 13

APF (alternate possibilities freedom). *See also* free will;
    PAP (principle of alternative possibilities)
    argument against, 168
    counterexample, 169 vs. determinism, 163–164
    and moral responsibility, 159–161
    overview, 158–159
Apology for Raymond Sebond, 62, 64
appeals
    to inappropriate authority fallacy, 43–44
    to the person fallacy, 42, 44–46
    to snobbery/vanity fallacy, 42–43
Aquinas, Thomas, 16, 190–192
arche ("first principle"), 12
arguments. *See also* claims; counterexamples; evidence
    from analogy, 36
    for APF (alternate possibilities freedom), 160–161
    cats example, 27–29
    circular, 46–47
    conditions of, 29
    deductive, 33
    defined, 25
    direct, 33–37
    disjunctive syllogism, 34–35
    enumerative induction, 35–36
    example-based, 33
    for existence of souls, 115
    inductive, 28, 35–37
    inductive generalization, 37
    inductive vs. deductive, 27
    inference to best explanation, 33
    modus ponens, 33–34
    modus tollens, 34
    purpose of, 32

arguments (*Cont.*):
    quality of, 29
    Republicans example, 28–29
    strategies, 32–33
    thought experiment, 33
    types of, 27
    using to evaluate claims, 25
    using to support claims, 33
    validity of, 27–29
arguments for skepticism, 70. *See also* skepticism
    Dream, 62–64
    Evil Genius, 64–66
    evil genius, 71
    Plato's Cave, 57–59
    unreliable sense experience, 60–62
Aristotle, 16
    death of, 16
    followers of, 54
    life span of, 218
    Virtue Theory, 218–219
artificial intelligence (AI), 127–129, 146
atheism, overview, 177–178
atheistic society, moral laws in, 253
Augustine, 16
authority, evaluating, 43–44
autonomous beings, defined, 256
auxiliary class, 246
axiomatic systems, examples of, 98
Ayer, 17

B

babies, behavior of, 83
Bacon, Francis, 7
Baillargeon, Reneé, 83
Bare Difference Argument, 39
Bealer, George, 95
begging the question fallacy, 46–47, 77–78
Behaviorism, 120–123
    failure of, 123
    Psychological vs. Philosophical, 121–122
beliefs
    basis of, 17–19
    justifying, 53–54
believing
    claims, 52–53
    senses, 60
Bennett, Jonathan, 199
Berkeley, George, 17, 99
Bhagavadgita, 6
Big Bang to rain (figure), 162
black box, mind as, 121, 123

blaming behavior, 165–166
Bloom, Paul, 83
body
    as condition for personhood, 136–137
    conscious mind in, 147–148
    identifying with, 145–146
    versus mind, 117
    in personal identity, 151–152
Bohm, David, 197
Bolzano, 17
BonJour, Laurence, 95
Braveheart, 157
de Broglie, Louis, 197

C

Calvin, John, 179
Campbell, George, 99
Carnap, Rudolph, 17, 103
Cartesian Circle, 78. *See also* Descartes, René
cat argument, 27–29
causality, law of, 95
causes, uncaused, 192
character, relevance of, 45
character-based moral theory, 217–221, 231
children, personhood of, 143
Chomsky, Noam, 122
Christ, message of, 13
circular argument, 46–47
circumstances, relevance of, 44–45
civil relationships, 144
claims. *See also* arguments; evidence
    about reality, 90
    ambiguity of, 76
    believing, 25, 52–53
    complex, 25–26
    counterexamples to, 37–39, 104
    descriptive vs. normative, 209
    disproving, 96
    evaluating truth of, 25
    evidence for, 30
    falsifiability of, 96–97
    knowing, 53
    knowing about logic, 94
    knowing about math, 94
    operators in, 25
    simple, 24
    supporting via arguments, 33
    true vs. false, 24–25
claims, content of, 19
Clement of Alexandria, 13
Clinton, Bill, 44–45

Cogito argument, 72–75, 78, 100
cognitive capacities, 135, 138
compatibilism
    argument against APF, 168
    intentions response, 170
    overview, 168–169
    PAP (principle of alternative possibilities), 169
    problems with, 171
complex claim, example of, 25–26
computers, comparing minds to, 128–129
Comte, August, 102–103
conclusion
    in cats argument, 27
    defined, 25
    following from premise, 29
    in Republicans argument, 28
    truth-value of, 29–30
Confucius, life span of, 218
Conscious Mind, 111–114, 130, 147–148. *See also* mind
consciousness, 152
consequence-based moral theory, 215–217, 220, 234
constitutional monarchy, 245
constitutional republic, 245
Constitutional rights, 245
Contextualism, 105
Copenhagen interpretation, 197
Copleston, Frederick, 188
cosmological argument, 189–197
counterexamples, 37–39. *See also* arguments
    Bare Difference Argument, 39
    to claims, 104
    Technique of Variant Cases, 39
criminal acts, view of, 168–169
The Critique of Pure Reason, 17

D

Darwin, Charles, 184–186
David, Michelangelo's statue of, 81–82
Davidson, 17
de Broglie, Louis, 197
decisions. *See also* moral claims; moral decisions
    as actions, 208
    clarifying terms, 236–237
    impermissible actions, 208–209
    influences on, 164
    moral vs. others, 208–209
    obligatory actions, 208
    permissible actions, 208–209
    supererogatory actions, 208
Declaration of Independence, 158, 247–248, 255–256
deductive arguments, 33–35

Deductivism, explained, 95–96
deontology, explained, 211
Descartes, René, 17, 63–66, 71. *See also* Cartesian Circle
    clear and distinct beliefs, 76–77
    Conscious Mind, 147–148
    four claims, 77
    knowledge of objects, 81–82
    vs. Locke, 79–80, 83
    mind versus body, 117
    pineal gland, 118
    substance dualism, 116–118
    three certain claims, 75–78
determinism, 161–164
    vs. APF (alternate possibilities freedom), 163–164
    evidence for, 164
dictatorship, 245
direct argument, 33–37
disjunctive syllogism, 34–35
Divine Command Theory, 237–239
divine right of kings, 247
Doctor Who, 137
dogmatists, defined, 54
Dream Argument, 62–63
dream argument vs. evil genius, 66
dreaming vs. waking life, 62–63
duck argument, 36
duty-based moral theory, 211–215, 220

E

Earth is flat hypothesis, 96–97
Eastern vs. Western philosophy, 4
economic relationships, 144
Einstein, Albert, 95, 102
Elea, Zeno of, 14
electron, concept of, 77
Eliminative Materialism, 124–125
empiricism
    Logical Positivism, 102–103
    after Logical Positivism, 103
    plausibility of, 80–85
    vs. rationalism, 79–83
empiricists, defined, 31, 78
Empiricus, Sextus, 54–56, 60
The Enlightenment, 116–117
enumerative induction, 35–36
Epicurus, 16
epistemology, 52, 207
    domain of, 9
    Internalism, 105
equality vs. fairness, 258

ethics. *See also* moral philosophy
    applied, 207–208
    bio, 208
    branches of, 207
    business, 208
    cyber, 208
    domain of, 10–11
    environmental, 208
    legal, 208
    metaethics, 207
    vs. morality, 206
    normative, 207
    sub-branches, 208
Euclidean Geometry, 78–79, 95
Euthyphro dialogue, 238
events, organization of, 94–95
evidence. *See also* arguments; claims
    defined, 30
    direct, 31
    example of, 30
    experiential, 31
    identifying types of, 31
    induction applied to, 97–98
    nonexperiential, 31
    objective, 177
    omission from knowledge, 53
    a posteriori, 94
    scientific, 31
    subjective, 177, 179
    for truth of claim, 30
    types of, 32
evidentialism
    agnosticism, 177–180
    atheism, 177
    vs. non-evidentialism, 177, 180
    theism, 177, 180
evil, problem of, 198–202
Evil Genius argument, 64–66
evil genius argument
    vs. dream argument, 66
    response to, 72–75
evolution, Darwin's theory of, 185–186
explanans, defined, 40
explanations
    competing, 41
    defined, 40
external freedom, 156
Externalism, 105

F

fairness vs. equality, 258
faith vs. reason, 19

fallacies. *See also* reasoning
    begging the question, 46–47, 77–78
    example of, 129
    formal vs. informal, 41–42
    genetic, 192–193
    straw man, 46
fallacies/appeals
    to inappropriate authority, 43–44
    to the person, 42, 44–45
    to snobbery/vanity, 42–43
fallacy of composition, 192–193
family relationships, 144
Fatalism, defined, 165
"first principle," water as, 12
Five Ways of Thomas Aquinas, 190–192
Frankfurt, Harry, 169–170
Freaky Friday, 148
free will, Kant's position on, 101–102. *See also* APF
        (alternate possibilities freedom)
freedom
    APF (alternate possibilities freedom), 158–159
    external, 156, 249
    internal, 156–157
    kinds of, 159
    and ownership, 249–251
    political, 157–158, 249
French Revolution, 158
Freud, Sigmund, 121
Functionalism, 127–129

G

Galileo
    writings on mechanics, 7
    writings on motion, 7
Gaunilo, 188
genetic fallacy, 192–193
genetics, Mendel's theory of, 184–185
Gettier, Edmund, 103–105
God
    concept of, 176–177
    Medieval Christian views of, 13
    traditional conception of, 199–200
God exists argument, 77–78
God's existence. *See also* religion
    agnosticism, 177, 179
    approaches to, 180
    arguments against, 198–199
    arguments for, 181–182
    atheism, 177
    considering evidence of, 182
    cosmological argument, 189–197
    evidential problem of suffering, 200–202

God's existence (*Cont.*):
    evidentialism, 177–180
    logical problem of suffering, 199–200
    non-evidentialism, 177, 180
    ontological argument, 187–189
    problem of evil, 198–202
    prudential arguments, 197–198
    prudentialism, 178–179
    questioning, 181, 202
    subjectivism, 179
    teleological argument, 182–187
    theism, 177
governments
    constitutional monarchy, 245
    constitutional republic, 245
    dictatorship, 245
    divine right of kings, 247
    freedom, 249–251
    laissez-faire, 247–248
    Liberalism, 245, 247–249
    oligarchy, 245
    ownership, 249–251
    paternal, 246–247
    Paternalism, 245
    reactive, 244
    theocracy, 245
    totalitarian regime, 245
Graham, Lindsey, 28
gravity, theory of, 102
"Greater Goods Defense," 200
Greek philosophers, position on souls, 116
Groundwork for the Metaphysics of Morals, 233

H

hard determinism, 165–166
hard indeterminism, 166–168
Hawking, Stephen, 136–137
Hegel, 17
Hobbes, Thomas, 247
Homer, 62
homosexuality, view of, 254
Hospers, John, 97–98
human action, motivations of, 164, 166–167
human behavior
    changing, 166
    explanation of, 164
human being, defining, 136
human experience, enriching, 5
human knowledge
    extent of, 9
    nature of, 9

humanity, views of, 245
Hume, David, 17, 89
    distinguishing impressions, 90
    vs. Hospers, 97–98
    human self, 152
    vs. Kant, 94, 100
    vs. Locke, 85, 90
    main argument, 92–93
    Problem of Induction, 91–93
    skepticism of science, 90–93
hypotheses
    falsifiability of, 96–97
    testing, 95–96

I

"I Think, therefore I am" argument, 72–75
idealism, explained, 99
ideas, obtaining from world, 83–84
"if and only if" operator, 26
"if..., then..." operator, 26
impermissible actions, 208–209
impressions, distinguishing, 90
indexicals, defined, 25
individual rights vs. common good, 254–257, 259–260
inductive arguments, 35–37. *See also* Problem of
    Induction
    applying to evidence, 97–98
    defined, 28
inference to best explanation, 33, 40–41
infinity, role in Kalam argument, 195–196
information, communicating, 142
intentions response, 170
internal freedom, 156–157
Internalism, 105
intrinsic vs. instrumental value, 212–213

J

James, William, 198
Jesus Christ, message of, 13
Johnny Get Your Gun, 136–137
Jung, Carl, 121
justice
    distributive, 258–259
    measure of, 258
    procedural, 258–259
justified true belief, 104

K

"Kalam Cosmological Argument," 193–197
Kant, Immanuel, 17–19, 99–102

Kant, Immanuel (*Cont.*):
    conditions for moral judgment, 101
    critique of moral theory, 214
    deontology, 211
    duty-based moral theory, 211–215
    eternal soul, 101
    existence of God, 101
    free will, 101
    Groundwork for the Metaphysics of Morals, 233
    vs. Hume, 94, 100
    intrinsic vs. instrumental value, 212–213
    life span of, 256
    moral theories, 230–232
    noumenal reality, 100
    phenomenal reality, 100–101
    transcendental philosophy, 100–102
killing arguments, 38–39
knowledge
    determining, 51–54
    justified true belief, 104
    necessary conditions, 52–53
    of objects, 81–82
    obtaining through experience, 99
    vs. probability, 80
    skepticism of, 54–58
    study of, 52
Kuhlmeier, Valerie, 83

L

laissez-faire government, 247–248
Lane Craig, William, 193–194
languages
    capacity for, 141–143
    components of, 122
    engaging in, 142
    rules of, 5
law
    moral vs. societal, 252
    and morality, 252–254
legal claims, early evaluation of, 7–8
Leibniz, 17, 79
Lewis, C. S., 238
liberal government, freedom of, 250
Liberalism, 245, 247–249
    classical, 248–249
    vs. Paternalism, 257
Locke, John, 17, 79–80, 148, 250
    vs. Descartes, 79–80, 83
    vs. Hume, 90
    ideas about objects, 82–83
    personal identity, 150
    reliance on probability, 85

logic, 207. *See also* operators
    domain of, 9
    knowing claims about, 94
    problem, 140
    rules of, 5
Logical Positivism, 102–103
Lucretius, 16
lunar eclipse, prediction by Thales, 12

M

major vs. minor premise, 188
Malebranche, 78
Marx, 17
Material Monism, 120
Mathematical Principles of Natural Philosophy, 7
mathematics, knowing claims about, 94
The Matrix, 65–66
medicine
    beginning of, 7
    as prescriptive enterprise, 7
medieval Philosophy, 16
Meditations on First Philosophy, 63–64
memories
    collection of, 149–150
    discontinuous, 150
    psychological continuity theory, 150
Mendel's theory of genetics, 184–185
mental states, capacity for, 141–143
metaphysical freedom. *See* APF (alternate possibilities freedom)
metaphysics, 207. *See also* reality
    domain of, 10
    influence of, 257
    and society in theocracy, 253
Michelangelo's statue of David, 81–82
Miletus, Thales of, 5, 12, 16
    as legend, 12–13
    philosophical approach, 12
    prediction of eclipse, 12
Mill, John Stuart, 17, 215, 248, 256
mind. *See also* Conscious Mind
    altering, 124–125
    ancient Greek conceptions, 115–116
    ancient Western conceptions, 114–115
    of animals, 111–113
    as black boxes, 121, 123
    versus bodies, 117
    comparing to computers, 128–129
    Conscious, 111–114
    distinguishing types of, 112–114
    identifying with, 147
    Perceptual, 111–114

mind (*Cont.*):
    qualities of, 110–111
    Reasoning, 111–114
    substance dualism, 116
mind-body problems
    Behaviorism, 120–123
    Functionalism, 127–129
    interaction, 119–120
    manifestation of, 119–120
    Material Monism, 120
    Mind-Brain Identity Theory, 123–126
Mind-Brain Identity Theory, 123–126
minor vs. major premise, 188
modus ponens argument, 33–34
modus tollens argument, 34
de Montaigne, Michel, 61–62, 64
Moore, 17
moral claims. *See also* decisions
    applying reasoning strategy, 232–234
    applying theories, 230–232
    clarifying terms, 226–227
    getting facts, 227–228
    identifying principles, 228–229
    making decisions, 235
moral communities, personhood in, 144–145
moral decisions
    consistency of, 210
    fairness of, 210
    making, 235
moral judgment, conditions for, 101
moral philosophy, 207. *See also* ethics
    normativity, 209–210
    overview, 205–208
moral prescriptive claims, 209–210
moral principles
    identifying, 228–230
    justifying, 252
moral reality, explained, 10
moral relativism, 236–237
moral responsibility, 143–144, 159–161
moral theories
    applying, 230–232
    character-based, 217–221, 231
    consequence-based, 215–217, 234
    duty-based, 211–215
    moral relativism, 236–237
    reductio ad absurdum, 235–236
    religious-based morality, 237–239
    testing, 233
moral vs. societal law, 252
morality. *See also* ethics
    vs. ethics, 206
    and law, 252–254

Moribus, 137
Moses Maimonides, 16
motion, theory of, 79
motivations, formation of, 164, 167
Müller-Lyer Illusion, 71–72

N

Nagel, Thomas, 130
natural selection, 185–186
nature vs. nurture, 164–165
necessary conditions
    for knowing, 52–53
    for personal identity, 151–152
Nietzsche, 17
New Instrument, 7
Newton, Isaac, 7, 79
normative claim, explained, 10
normativity
    overview, 209–210
    rationality, 210
"not" operator, 26
Novum Organum, 7

O

Obama, Barack, 44–45
objective evidence, defined, 177
objects
    observing, 121
    obtaining, 84
    primary qualities, 83–84
    properties of, 83
    secondary qualities, 83–84
obligatory actions, 208–209
The Old Farmer's Almanac, 5–6
oligarchy, 245
ontological argument, 187–189
operators. *See also* logic
    "and," 26
    in claims, 25
    "if and only if," 26
    "if..., then...," 26
    indexicals, 25
    modal, 25
    "not," 26
    "or," 26
    quantifiers, 25
    types of, 25
"or" operator
    in complex claim, 26
    in disjunctive syllogism, 34–35

Origin of Species, The, 184
ownership and freedom, 249–251

P

pain, problem of, 198–202
Paley, William, 183–184
PAP (principle of alternative possibilities), 169. *See also* APF (alternate possibilities freedom)
paradox
    defined, 14
    example of, 14–15
Pascal, Blaise, 19, 198
paternal governments, restriction of freedoms in, 250–251
Paternalism, 245–247
    argument for, 249
    vs. Liberalism, 257
    motivation for, 251
Pavlov, Ivan, 121
pen, dropping, 90–91
perceptions
    distinguishing, 90
    theory of, 84–85
Perceptual Mind, 111–114
permissible actions, 208–209
persistent vegetative state (PVS), 145, 147
personal identity
    body, 145–146, 151
    memories, 149–150
    mind, 147, 151
    necessary conditions, 151–152
    sufficient conditions, 151–152
personhood
    body as condition for, 136
    capacity for reason, 139–140
    of children, 143
    cognitive capacities, 135
    criteria, 134–136
    determining, 134, 137–139
    language, 141–143
    mental states, 141–143
    in moral communities, 144–145
    moral responsibility, 143–144
    rationality, 139
    social relationships, 143–144
perspective, examples of, 61
Philo, 16
philosophers. *See also* thinkers
    al-Ghazali, 16, 193
    Anaximander, 16
    Aquinas, Thomas, 16

philosophers (*Cont.*):
    Aristotle, 16, 218
    Augustine, 16
    Ayer, 17
    Berkeley, 17
    Bolzano, 17
    Campbell, George, 99
    Carnap, Rudolph, 17, 103
    Comte, August, 102–103
    concerns of, 6
    Confucius, 218
    Copleston, Frederick, 188
    Davidson, 17
    Descartes, René, 17, 63–66, 71, 75–78
    empiricists, 31
    Empiricus, Sextus, 54–56, 60
    Epicurus, 16
    Gettier, Edmund, 103–105
    Hegel, 17
    Hospers, John, 97–98
    Hume, David, 17, 85
    Kant, Immanuel, 17–19, 99–102, 230–232, 256
    Leibniz, 17, 79
    Locke, John, 17, 79–80
    Lucretius, 16
    Marx, 17
    Mill, John Stuart, 17, 215, 248, 256
    de Montaigne, Michel, 61–62, 64
    Moore, 17
    Moses Maimonides, 16
    Nietzsche, 17
    objective reality, 4–5
    Pascal, Blaise, 19, 198
    Philo, 16
    Plantinga, Alvin, 189
    Plato, 16, 53, 62–63, 115, 218, 238, 246
    Pollock, John, 75
    Popper, Karl, 95–96
    rationalists, 31
    Reid, Thomas, 99
    Rousseau, 17
    Rowe, William, 179
    Russell, Bertrand, 7, 17–18, 74
    Salmon, Wesley, 95
    Smith, Adam, 248
    Socrates, 16, 56
    Spinoza, 17, 78–79
    St. Anselm, 16
    St. Augustine, 16
    Stoics, 16
    Thales of Miletus, 5, 12–13, 16

philosophers (*Cont.*):
   Wittgenstein, 17
   Xenophon, 16
   Zeno of Elea, 14, 16
Philosophiae Naturalis Principia Mathematica, 7
Philosophical Behaviorism, 121–122
philosophical domains
   epistemology, 9
   ethics, 10–11
   logic, 9
   metaphysics, 10
   overview, 8
   political, 11
philosophy
   branches of, 207
   changes in approach, 11–12
   components, 4
   contemporary, 17
   defined, 4
   Eastern vs. Western, 4
   epistemology, 52, 207
   fields of, 6
   history of, 16–17
   logic, 207
   medieval, 16
   metaphysics, 207
   modern, 17
   moral, 207
   Plato, 12
   political, 207, 244–246
   practical vs. theoretical, 8
   pre-Socratic era, 16
   Roman, 16
   vs. science, 6–7
   skill-based aspect of, 17
   Socratics, 16
   subjects of interest, 11–12
   value of, 17–19
   Xenophanes, 13
physicians, early practices of, 7
physics
   explanations of, 233
   theory of, 7
   theory of motion, 79
pineal gland, Descartes and, 118
Plantinga, Alvin, 189
Plato, 16, 62–63, 115
   auxiliary class, 246
   class structure, 246
   commentary about Thales, 12
   Euthyphro dialogue, 238
   followers of, 54

Plato (*Cont.*):
   life span of, 218
   producer class, 246
   ruling class, 246
   Theatetus dialogue, 53, 62–63
Platonism, 56
Plato's Cave argument, 58–59
political, 207
political freedom, 157–158, 249
political issues, evaluating, 260
political philosophy, 11, 244–246
Pollock, John, 75
Popper, Karl, 95–96
practical philosophy, 8
pragmatic studies, 5–6
praising behavior, 165–166
premise
   in cats argument, 27
   defined, 25
   following conclusion from, 29
   major vs. minor, 188
   in Republicans argument, 28
Pre-Socratics, 16
principle of alternative possibilities (PAP), 169. *See also*
   alternate possibilities freedom (APF)
probability
   forward-looking, 97
   vs. knowledge, 80
Problem of Induction. *See also* inductive arguments
   avoiding, 95–96
   disproving claims, 96
   introduction by Hume, 91–93
   justification of, 95
   a posteriori claims, 94
   a priori claims, 93–95
   probability, 97
   solving, 97–98
Process Reliabilism, 105
producer class, 246
proofs of Aquinas, 190–192
property, ownership of, 250–251
property dualism vs. substance dualism, 119–120
prudentialism
   arguments, 197–198
   overview, 178–179
Psychological Behaviorism, 121
psychological continuity theory, 150
psychological freedom, 156–157
PVS (persistent vegetative state), 145, 147
pyrrhic, defined, 55
Pyrrhic victory, 55

Pyrrhonian skepticism, 54–55
    Descartes' response to, 72–73
    refuting, 70–71
Pyrrhonism vs. Academic skepticism, 57–58

Q

quantifiers, examples of, 25
quantum fluctuation, 197
quantum physics, development of, 196–197

R

radioactive decay, 196–197
rational interference, principles of, 9
rationalism vs. empiricism, 79–83
rationalists, defined, 31, 78
rationality
    defined, 139
    overview, 210
reality. *See also* metaphysics
    claims about, 90
    forming beliefs about, 17–19
    immediate perceptions of, 71
    moral, 10
    nature of, 10
    nonreligious aspect of, 12
    noumenal, 100
    objective, 4–5
    per Thales of Miletus, 12
    phenomenal, 100–101
    pragmatic aspect of, 12
    principled aspect of, 12
    religious studies of, 6
    studying systematically, 4–5
    systematic study of, 4
reason vs. faith, 19
reasoning. *See also* fallacies
    capacity for, 139–140
    error in, 41
    use of, 5
Reasoning Mind, 111–114
reductio ad absurdum, 235–236
The Reformation, 16
Regarding Henry, 149, 152
Reid, Thomas, 99
relationships, types of, 144
religion. *See also* God's existence
    impact on study of reality, 6
    philosophy of, 176
religious-based morality, 237–239
The Representation Theory of Perception, 84

The Republic, 56, 58
Republicans argument, 28–29
rights
    vs. common good, 254–257
    defined, 254
    of individuals, 259–260
robots, potential of, 137–138, 142–143
Roman Philosophy, 16
Rousseau, 17
Rowe, William, 179–180
ruling class, 246
Russell, Bertrand, 7, 17–18, 74

S

Salmon, Wesley, 95
Schiavo, Terry, 145–147, 151
science
    fields of, 7
    origin of terminology, 7
    vs. philosophy, 6–7
    skepticism of, 90–93
scientific method, development of, 7
Scientific Revolution, 120
senses, reliability of, 60–62
sensory beliefs, doubting, 65
simple claim, 24–25
skepticism. *See also* arguments for skepticism
    Academic, 54
    defined, 54
    after Medieval period, 116–117
    Pyrrhonian, 54
    of science, 90–93
skepticism responses
    Cogito argument, 72–75, 78
    Descartes, René, 70–72
    Müller-Lyer Illusion, 71–72
skeptics. *See* Hume, David
Skinner, B.F., 121
Smith, Adam, 248
social relationships, 143–144
societal vs. moral law, 252
Socrates, 16, 56
Socratic Method, 55
solipsism, 73, 78, 85
Song of the Lord, 6
soul, arguing existence of, 115–116
Spelke, Elizabeth, 83
Spinoza, 17, 78–79
St. Anselm, 16
St. Augustine, 16
Stoics, 16

straw man fallacy, 46
subjective evidence, defined, 177
subjectivism, overview, 179
substance dualism
    Descartes's argument, 116–118
    explained, 116
    vs. property dualism, 119–120
    scientific perspective, 118
suffering
    evidential problem of, 200–202
    logical problem of, 199–200
sufficient conditions for personal identity,
        151–152
supererogatory actions, 208–209
syllogism, defined, 34–35

T

tabula rasa, 80
Technique of Variant Cases, 39
teleological argument, 182–187
Thales of Miletus, 5, 12, 16
    as legend, 12–13
    philosophical approach, 12
    prediction of eclipse, 12
Theatetus dialogue, 53, 62–63
theism, overview, 177, 180
theocracy, 245, 253
theoretical philosophy, 8
theoretical virtues, explained, 40–41
theories, inadequacy of, 233
Theory of Relativity, 95, 102
thinkers, concerns of, 5. *See also* philosophers
Thomas Aquinas's Five Ways, 190–192. *See also*
        Aquinas, Thomas
thought experiment, 33
    counterexample, 37–39
    defined, 37
Tolman, Edward, 123
totalitarian regime, 245
transcendental philosophy, 100
Trojan War, 55

truth vs. usefulness, 5–6
truth-value, evaluating for conclusion, 29–30

U

universal grammar, 122–123
U.S. Declaration of Independence, 158, 247–248,
        255–256
utilitarianism, 215–220

V

validity
    definitions, 27–28
    meanings of, 37
vices vs. virtues, 219
"Vienna Circle," 103
views
    explaining vs. accepting, 59
    explaining vs. defending, 57
Virtue Theory, 218–221

W

Wager Argument, 198
Wallace, William, 157–158, 249
Watchmaker Argument, 183–184
water, as first principle, 12
Watson, John, 121
wax example, 81–83
Western vs. Eastern philosophy, 4
"The Will to Believe," 198
Wittgenstein, 17
Wynn, Karen, 83

X

Xenophanes, 13
Xenophon, 16

Z

Zeno of Elea, 14, 16